THE QUANTUM CONSCIOUSNESS

Navigating the Unseen Realms of Reality and Self

Pietro Paolo

ISBN: 9798345474747
Cover design by Pietro Paolo
Printed in USA

Table of Contents

Introduction to the Infinite Complexity of Existence

The Pursuit of Knowledge Beyond Boundaries

The drive to explore the unknown, to push beyond the comfortable edges of knowledge, has fueled human progress since the dawn of awareness. Yet, beyond the boundaries of established understanding lies a landscape of complexity so vast, so intricate, that it can feel unknowable. This realm isn't bound by linear paths or simple answers—it defies traditional frameworks and invites those who seek to question everything they believe to be true. This book is an invitation into that realm, where knowledge is not merely accumulated but fundamentally transformed, and where truth itself may be a shifting, multi-dimensional concept.

What lies beyond the boundary of conventional knowledge is not just a deeper understanding of the universe but a profound reshaping of the self. The journey into this complexity demands a readiness to dissolve preconceived notions, to abandon the rigid constructs that define the self, reality, and even time. Here, the familiar rules of cause and effect, objective and subjective, inner and outer, no longer hold as separate entities. Instead, they merge into a dynamic interplay that requires a mind agile enough to think beyond dualities, a mind prepared to experience both certainty and mystery as dual expressions of the same underlying truth.

The Nature of Boundaries

Boundaries in human thought often serve to comfort, to categorize, and to protect us from the unsettling infinity that surrounds and permeates all of existence. They provide containment, shaping knowledge into something that feels achievable, understandable, and actionable. But boundaries are, by nature, limitations. They restrict not only what we are willing to accept as true but also what we are capable of imagining. To transcend these boundaries is to let go of this need for containment, to surrender to the mystery that lies just beyond the reach of words.

The pursuit of knowledge beyond boundaries requires a radical approach, one that does not seek to conquer or solve, but rather to participate in a more intimate, resonant relationship with the unknown. Knowledge in this space is not acquired; it is revealed through experience, intuition, and an open willingness to confront paradox. It is a knowledge that lives and breathes, one that grows and shifts with each new insight, each deepened awareness.

A Call to the Seeker of Truth

For those who seek to venture beyond what is known, the journey is both daunting and exhilarating. There are no maps, no final answers, only glimpses of a truth that is vast and mutable. Here, each discovery opens a new set of questions, each insight hints at deeper mysteries. To pursue this knowledge is to accept the unknown as a partner, to invite uncertainty as a companion, and to surrender to the notion that understanding is an ever-evolving process without a final destination.

In this pursuit, the lines between science and philosophy, objective and subjective, real and imaginary, begin to blur. The seeker learns that all boundaries are ultimately constructs, temporary definitions that help us organize our reality but that do not contain or limit the essence of reality itself. In stepping beyond these constructs, one realizes that true knowledge exists not within the isolated silos of disciplines, but in the interconnectedness of all things—a web so intricate that each thread vibrates with the resonance of the whole.

Beyond the Ego, Beyond the Known

This journey into infinite complexity is not just intellectual; it is transformative. It requires letting go of the ego's attachment to certainty, to identity, and to control. The mind must become pliable, flexible enough to expand and contract with the flow of knowledge, aware that each new discovery is merely a stepping stone to the next horizon. Here, the seeker does not conquer truth but is transformed by it. Knowledge becomes a dance, an exchange where the self is reshaped with each step.

In the chapters that follow, we will delve deeper into the fabric of this complexity, exploring consciousness, reality, self, and the paradoxes that lie at the intersection of what we know and what we can only imagine. This is not a journey for those who seek comfort or certainty; it is for those who are willing to question, to wonder, and to embrace the mystery that lies at the heart of existence.

This book, then, is a guide—a map that is, paradoxically, incomplete, for no map can truly chart the terrain of the infinite. It is an invitation to think, feel, and experience beyond the edges of what you know, to step into the vastness of possibility that awaits. The pursuit of knowledge beyond boundaries is not about finding answers; it is about expanding the questions, about experiencing the richness and beauty of existence in all its unfathomable complexity. Welcome to the journey.

Why This Book Is Not for Everyone (and Why It's Essential for Some)

This book is a threshold—a doorway into realms of thought and experience that require more than simple curiosity. It demands an open mind, a resilient heart, and a willingness to embrace discomfort. It is a journey into the unknown, a quest that does not yield clear answers or tidy conclusions but instead challenges its readers to question everything they think they know. For some, this may feel like stepping off solid ground into an ocean of uncertainty, a disorienting plunge into ideas that destabilize the familiar comforts of life. This book is not intended for everyone because it asks for commitment, courage, and a suspension of the usual ways of understanding the world.

Not everyone is ready or willing to take this journey. Many people are content within the structures of conventional knowledge, within the rules and boundaries that define daily life. For them, mystery is something to be avoided or explained away, and uncertainty is uncomfortable, even intolerable. The quest to understand consciousness and the fabric of reality requires stepping beyond such limitations. This journey requires a type of seeing that is not focused on finding final answers but on embracing continuous inquiry. For those who thrive on concrete truths, absolutes, and certainty, this book may feel like a paradoxical exercise, a book that offers more questions than answers.

A Path for the Seeker

Yet, there exists a small group—a less-than-1%—who feel an unrelenting drive to look deeper, to venture further, to explore beyond what is easily seen. For these individuals, the ordinary explanations of life feel insufficient. They are seekers, those who are willing to leave behind the familiar landscapes of understanding and dive into the depths of the unknown. They sense that beneath the visible layers of life lies an intricate web of truths, woven with threads of mystery and meaning that are only accessible to those who are unafraid of the dark. This book is for them.

For these seekers, the complexity of existence is not something to be simplified but rather something to be engaged with fully. They understand that knowledge itself is a living, breathing entity, one that evolves and expands with each new insight. To them, consciousness is not just a byproduct of neural activity but a profound mystery that holds keys to understanding reality itself. They are drawn to the unknown, not because it provides certainty, but because it offers a path to a deeper understanding of themselves and the cosmos. For them, this book is essential, for it offers a companion in their quest, a mirror to reflect their own thoughts, doubts, and revelations.

Why This Journey Matters

If you are one of these seekers, then this book is essential, not because it holds ultimate truths but because it opens doors—doors that lead inward and outward simultaneously, to the vast reaches of both the cosmos and the self. You will not find clear directions or easily packaged wisdom here. Instead, you will find pathways that will lead you to ask deeper questions, to wonder, and to imagine. You will find perspectives that may unsettle you, that may lead you to abandon previously held beliefs in favor of a more expansive view of existence.

This journey is crucial because understanding the nature of consciousness and reality affects how we live, how we relate to others, and how we view our place in the universe. For those willing to question, to risk their comfortable notions of self and world, the rewards are profound. This journey fosters not only intellectual growth but an expansion of empathy, humility, and awe. It connects the seeker to the world in a way that transcends the superficial, offering a sense of unity with all that is.

The Transformative Power of the Unknown

For those who dare to read on, this book will serve as a catalyst. It will challenge you to look within, to confront the unknown not as something external but as an intrinsic part of who you are. Here, you are invited to see the boundaries of knowledge not as barriers but as portals to deeper awareness. By embracing what is complex, what is mysterious, you may find yourself transformed, your perspective on reality reshaped in ways that you cannot yet imagine.

So, let this be a guide, a companion for those rare individuals who are willing to engage with life on its deepest levels. If you find yourself here, reading these words, then perhaps you are one of those few. This book may be difficult, it may unsettle and challenge, but if you are ready, it has the potential to offer something invaluable—a glimpse into the infinite complexity of existence and the transformative power of understanding it.

Chapter One: The Fundamental Paradoxes of Reality

Unifying Opposites: The Nature of Paradox as Universal Language

Reality, in its essence, is woven from contradictions. At the deepest levels, paradox is not a flaw but a feature—an integral characteristic of existence that defies our need for simple, linear understanding. This nature of paradox transcends cultures, philosophies, and sciences, revealing itself as a universal language that speaks to the profound complexity underlying all things. It invites us to see that opposites are not isolated or irreconcilable; rather, they are complementary aspects of a whole, dynamic and alive, constantly interacting in an endless dance.

Paradox is where logic reaches its limits and intuition begins. It beckons us to expand our thinking, to hold opposing truths together without forcing one to dominate or invalidate the other. In this way, paradox is not a puzzle to solve but a reality to experience—a language that speaks directly to the nature of consciousness, existence, and the self. Understanding paradox requires us to become comfortable with ambiguity, to cultivate a mind flexible enough to embrace contradiction without seeking to resolve it. When we learn to see through the lens of paradox, we are drawn into a more intimate relationship with reality itself, one that reveals the interconnectedness of all things.

The Duality of Light and Darkness

One of the most fundamental paradoxes that exists in the natural world is the duality of light and darkness. Light and darkness are often perceived as opposites, forces in opposition, where one must conquer the other.

14

However, without darkness, light has no context, no way to express itself fully. Darkness is not the absence of light but a complementary state that allows light to be seen, understood, and felt. Together, they form a continuum, creating a dance that shapes our perceptions of space, time, and form. This paradox serves as a metaphor for many of life's mysteries: joy and sorrow, creation and destruction, presence and absence.

The language of paradox teaches us that to truly understand light, we must also understand darkness, to recognize that one cannot exist without the other. In every moment of illumination lies a shadow, and in every shadow, there is the potential for light. Paradox here serves as a reminder of the wholeness of existence, a framework where seemingly opposite states are bound together in a relationship that defies the need for linear logic. It calls us to accept that life itself is an interplay of contrasts, a balance that gives meaning to our experiences and depth to our understanding.

The Quantum Paradox: Particle and Wave Duality

In the realm of quantum mechanics, the particle-wave duality is another vivid illustration of paradox as a universal language. Particles such as electrons or photons exhibit both wave-like and particle-like properties, depending on the conditions of observation. When measured in one way, they behave like particles, distinct and localized. When measured in another, they spread out like waves, existing in multiple places simultaneously. This phenomenon, known as wave-particle duality, defies the conventional logic that an object should be one thing or another— particle or wave—but not both.

This quantum paradox reveals that our reality is not confined to one rigid structure; it is fluid, shifting according to perspective. The same quantum entity can be seen as discrete or as continuous, as limited or as infinite. In this context, paradox becomes a portal to understanding the fundamental fluidity of the universe, a language that encourages us to embrace multiple perspectives simultaneously. Wave-particle duality is not just a quirky phenomenon in physics; it is a profound insight into the flexible nature of reality itself. To comprehend it is to glimpse the interconnectedness of all things and to appreciate that existence is far more nuanced than our categories allow.

The Self and the Other: A Paradox of Identity

In our own consciousness, we encounter paradox in the form of identity—the relationship between the self and the other. We experience ourselves as distinct individuals, separate from those around us, yet at the same time, we are intrinsically connected to others, to nature, to all of existence. This paradox of individuality and unity is a recurring theme in both spiritual traditions and modern science, pointing toward a deeper truth that transcends separation. We are, in essence, both unique and universal, isolated and interconnected, solitary and yet part of a vast network of consciousness.

The self-other paradox challenges our understanding of identity, pushing us to recognize that our sense of self is fluid, malleable, and ever-changing. It is shaped by interactions, relationships, and shared experiences, revealing that identity is not an isolated construct but a relational one. To understand ourselves fully, we must see ourselves within the context of the whole. Paradoxically, the more we recognize our interconnectedness, the more distinct and authentic our individual self becomes. By embracing this paradox, we move closer to a sense of unity, a realization that individuality and oneness are not opposing forces but complementary aspects of existence.

Paradox as the Language of Consciousness

The presence of paradox in every aspect of life hints at the nature of consciousness itself. Consciousness is not a static phenomenon but a dynamic field that interacts with the complexities of reality, constantly processing dualities and contradictions. It is through this engagement with paradox that consciousness evolves, that understanding deepens. By holding opposites together, consciousness transcends the limitations of linear thought, moving into a realm where understanding is intuitive, holistic, and expansive.

In this sense, paradox is the native language of consciousness, the medium through which we communicate with the deeper aspects of reality. It invites us to expand our perceptions, to cultivate a way of thinking that does not shy away from ambiguity but embraces it. To engage with paradox is to train the mind to think beyond binary choices, to explore possibilities that exist outside traditional categories, and to reach for insights that are profound precisely because they cannot be easily explained.

16

Living with Paradox: A Path to Wisdom

Embracing paradox is not just an intellectual exercise; it is a path to wisdom. In accepting the coexistence of opposites, we learn to navigate life's uncertainties with grace and insight. We develop resilience, adaptability, and a deeper appreciation for the complexity of existence. By seeing the unity within duality, we gain a more holistic understanding of ourselves and the world around us. We come to realize that true wisdom lies not in the certainty of answers but in the depth of questions, in the willingness to remain open to mystery, and in the courage to accept that some truths may forever remain beyond our grasp.

In learning to live with paradox, we find freedom—freedom from the need to categorize, to judge, to confine life within rigid definitions. Paradox teaches us to see the world as a continuous, unfolding interplay of forces that both oppose and complement each other. It allows us to approach reality with humility, knowing that no matter how much we learn, there is always more to discover. In this way, paradox becomes a guide, a wise and patient teacher that leads us beyond the superficial and into the profound, where we may find a greater truth, one that encompasses both light and darkness, unity and diversity, the self and the other.

This is the invitation of paradox: to think beyond the known, to experience beyond the familiar, and to understand that the ultimate language of the universe is not one of separation but of integration, a unifying dance of opposites that reveals the essence of reality itself.

Exploring the Role of Uncertainty and Ambiguity

Uncertainty and ambiguity are often viewed as impediments to knowledge, sources of frustration or discomfort that should be minimized or eliminated. Yet, within the vast framework of existence, uncertainty and ambiguity are not obstacles but essential elements of reality itself. They are the spaces between, the fertile grounds where possibility emerges, where rigid lines blur, and where truth reveals itself as layered, fluid, and multidimensional. Uncertainty and ambiguity invite us to move beyond the confines of black-and-white thinking, to embrace a world that is, at its core, a tapestry of overlapping shades, complexities, and contradictions.

In this chapter, we explore uncertainty and ambiguity not as adversaries to clarity but as portals to a deeper, more comprehensive understanding of the universe and ourselves. They challenge the mind to let go of its need for control, for absolutes, and to engage with reality in a way that is open-ended and adaptable. In doing so, we unlock new dimensions of consciousness, becoming more attuned to the nuanced and ever-evolving nature of existence.

Uncertainty: The Foundation of Growth and Discovery

Uncertainty is not simply the absence of knowledge; it is the fertile ground from which all knowledge springs. If everything were known, if every question had a definitive answer, there would be no room for growth, no space for discovery. The unknown is the canvas upon which we paint our understanding of reality, and uncertainty is the brush that allows us to navigate it. In physics, for instance, Heisenberg's uncertainty principle reveals that at the quantum level, the exact position and momentum of a particle cannot be simultaneously known with absolute precision. This isn't a limitation of measurement but a fundamental property of nature, suggesting that reality itself possesses an inherent indeterminacy.

This quantum uncertainty is mirrored in the macrocosm of human life. We live each day without complete knowledge of what the future holds, and it is this very uncertainty that fuels our curiosity, our creativity, and our drive to explore. Uncertainty allows for freedom, for the spontaneous emergence of new ideas and possibilities. In embracing uncertainty, we open ourselves to transformation, to the continuous unfolding of life in ways that may challenge our expectations but ultimately enrich our experience.

When we accept that uncertainty is intrinsic to existence, we begin to see it not as something to overcome but as something to embrace. It invites us to approach life with a sense of wonder, to recognize that the unknown holds a beauty and potential that certainty could never offer. Each moment becomes an invitation to engage with the world in a state of openness, where discovery is a process without end.

Ambiguity: The Art of Holding Multiple Truths

Ambiguity, like uncertainty, defies simple categorization. It is the recognition that a single event, idea, or experience can contain multiple, even conflicting, interpretations. In a world that often seeks definitive answers, ambiguity challenges us to hold space for complexity, to acknowledge that truth is not always singular or straightforward. In fact, some of the most profound truths are inherently ambiguous, inviting us to look beyond the surface and to seek meaning in the interplay of multiple perspectives.

Ambiguity requires the mind to operate beyond binary thinking, to develop a tolerance for gray areas where clear answers do not readily emerge. For instance, in relationships, emotions are rarely clear-cut; love and resentment, hope and fear, joy and sorrow can coexist, creating a nuanced experience that cannot be reduced to simple definitions. Similarly, in art, literature, and philosophy, ambiguity adds depth and richness, allowing for interpretations that resonate differently with each individual. The ambiguous aspects of a piece of art, for example, can evoke a myriad of personal insights and emotions, drawing each observer into their own unique experience of truth.

By learning to hold multiple truths simultaneously, we develop a flexible and expansive mind, one that can move between perspectives and see the interconnectedness of ideas that might initially seem contradictory. Ambiguity teaches us that life is not always meant to be solved like a puzzle but to be appreciated as a complex, evolving story. It encourages us to engage with the world in a way that is more open, more compassionate, and more inclusive, as we begin to understand that every viewpoint has its own validity, its own piece of the greater whole.

The Dance of Uncertainty and Ambiguity in Consciousness

In the realm of consciousness, uncertainty and ambiguity are not merely abstract concepts; they are dynamic processes that shape our understanding and perception of reality. Consciousness itself thrives on these qualities, for they create the space in which new ideas can emerge, where insights can be born out of the interaction between the known and the unknown. When we allow uncertainty and ambiguity to coexist within us, our consciousness becomes like a river—fluid, adaptable, and capable of flowing into new channels of thought.

Consciousness, then, is not a static state but a continuous dance with the unknown. It is through this dance that we expand our awareness, reaching beyond the limitations of fixed beliefs and rigid interpretations. By allowing ourselves to live in the tension of ambiguity and the openness of uncertainty, we cultivate a form of wisdom that is not reliant on definitive answers but is rooted in a deep appreciation for the mystery of existence. This wisdom understands that knowledge is not a destination but a journey, one that is enriched by every question that remains unanswered.

The Transformative Power of Embracing the Unknown

When we embrace uncertainty and ambiguity, we allow ourselves to be transformed. We release the need for control and embrace a state of mind that is more receptive, more attuned to the nuances of reality. In this state, we become explorers of consciousness, willing to venture into realms where clear answers are elusive but profound insights await. We begin to see that ambiguity is not a flaw but a feature of a universe that is far richer and more intricate than we can fully comprehend.

This willingness to embrace the unknown is also a source of resilience. Life, with all its unpredictability, requires a capacity to adapt, to let go of certainty in favor of growth. Those who can navigate the world with a mind open to uncertainty are better equipped to face challenges, to move through change with grace, and to find meaning in circumstances that others might find disorienting. The acceptance of uncertainty fosters a strength rooted not in unyielding belief but in the ability to evolve, to learn, and to discover continuously.

Embracing Uncertainty and Ambiguity as a Path to Wholeness

In a world that often prizes clarity and certainty, the ability to embrace uncertainty and ambiguity is a rare and valuable skill. It opens us to a fuller, more nuanced understanding of reality and allows us to engage with life in a way that is deeply enriching. By making peace with the unknown, we discover a sense of wholeness—a feeling of connection to the vast, interconnected web of existence that transcends our individual experiences.

This journey into the heart of uncertainty and ambiguity is ultimately a path to self-discovery. As we venture into the unknown, we come to see that the boundaries we once held between self and other, between reality and perception, begin to dissolve. We realize that we are not isolated observers but active participants in the unfolding mystery of existence, and that our understanding of reality is, in itself, a living process.

In this way, uncertainty and ambiguity become powerful allies, guiding us toward a deeper sense of truth and unity. They remind us that the universe is an infinite realm of possibilities, one that reveals itself only to those who are willing to look beyond the surface, to question deeply, and to embrace the beauty of not knowing.

Chapter Two: Consciousness as a Quantum State

Consciousness and Quantum Mechanics: The Bridge between Observer and Observed

In the quantum realm, where particles can exist in multiple states simultaneously and where reality is probabilistic rather than deterministic, the role of the observer becomes more than just passive. Observers influence outcomes simply by observing. This phenomenon—often described as the "observer effect"—raises profound questions about the nature of reality, suggesting that consciousness itself may be an active participant in shaping the physical world. In this context, consciousness is not merely an internal experience; it becomes a bridge between the observer and the observed, binding them in an inseparable relationship where each influences the other.

The idea that consciousness affects reality, that it participates in the unfolding of events at the quantum level, disrupts traditional scientific models that separate the mind from the physical world. Quantum mechanics reveals a universe in which subjectivity and objectivity are intertwined, a universe that defies strict separation between the observer and the phenomenon being observed. The role of consciousness in this framework is not to passively witness reality but to engage with it, to influence it in subtle yet profound ways. This bridge between observer and observed lies at the heart of a new understanding of reality—one where consciousness is not merely a byproduct of matter but an integral part of existence itself.

The Double-Slit Experiment: An Illustration of the Observer Effect

One of the most well-known demonstrations of the observer effect is the double-slit experiment, a foundational experiment in quantum mechanics. When particles such as electrons are shot through two slits, they exhibit wave-like behavior, creating an interference pattern on a detection screen that suggests they travel through both slits simultaneously. However, when the particles are observed or measured as they pass through the slits, this wave-like behavior collapses, and the particles behave as if they went through only one slit, creating a pattern consistent with particle-like behavior.

This experiment implies that the act of observation influences the behavior of particles, causing them to adopt either a wave or particle form based on whether they are being observed. The interference pattern disappears when an observer is introduced, suggesting that the conscious act of measurement affects the fundamental nature of the particles. This raises profound questions: How can mere observation alter the state of physical reality? What does it mean for the observer to play an active role in determining the nature of existence?

The double-slit experiment reveals that at the quantum level, the boundary between observer and observed is permeable, dynamic, and inseparable. Consciousness, in this context, becomes an agent of transformation, a catalyst that determines which potential states become manifest. This phenomenon suggests that reality is not a fixed structure waiting to be discovered but a fluid field of potential that is shaped, at least in part, by the presence of a conscious observer.

Quantum Superposition and the Role of Consciousness

Quantum superposition—the ability of particles to exist in multiple states simultaneously—is another phenomenon that challenges traditional concepts of reality and highlights the role of consciousness. In a superposition, a particle does not have a single, definite state; instead, it exists in a probabilistic combination of all possible states until observed. When measured, however, the particle "collapses" into one specific state, as if the conscious act of observation determines which of the potential outcomes becomes real.

This raises an intriguing possibility: that consciousness is, in some way, responsible for bringing potential into actuality. Without observation, quantum particles exist in a state of possibility, a haze of probabilities

23

where every potential reality coexists. It is only when consciousness interacts with this field that one particular outcome crystallizes. This suggests a universe in which reality is not fixed but is shaped by the interplay of consciousness and potential—a co-creative dance where the mind participates in the formation of the world around it.

Superposition also raises questions about the nature of time and causality. If particles exist in multiple states until observed, then reality itself is not fully determined until consciousness engages with it. This idea aligns with the concept of a participatory universe, a reality that is not static but is continuously shaped by the act of observation, by the bridge between the conscious observer and the quantum world.

The Quantum Mind Hypothesis: Consciousness as a Quantum Phenomenon

Some scientists and philosophers propose that consciousness itself may operate according to quantum principles, an idea known as the quantum mind hypothesis. This hypothesis suggests that the brain may function in a way that is analogous to quantum systems, where consciousness arises from the superposition, entanglement, and coherence of neural processes. In this view, consciousness is not a byproduct of purely classical physical processes but a quantum phenomenon that reflects the same uncertainty, ambiguity, and interconnectedness found in the quantum world.

The quantum mind hypothesis posits that consciousness may emerge from quantum processes within the brain, that our thoughts, emotions, and perceptions may be influenced by the same principles that govern the behavior of subatomic particles. This would mean that consciousness itself is inherently uncertain, fluid, and interconnected with the larger universe in ways we are only beginning to understand. If consciousness operates on quantum principles, it would naturally possess the qualities of uncertainty and superposition, existing in multiple states and constantly interacting with potential realities. This would explain why consciousness can feel expansive, boundless, and capable of reaching beyond the immediate sensory experience.

Such a hypothesis suggests that consciousness does not merely observe reality but actively participates in it, in the same way that particles interact and influence one another in quantum entanglement. In this view, the bridge between observer and observed is not a metaphor but a literal connection, a dynamic interplay where consciousness and the physical world are entangled, shaping each other in a continuous feedback loop.

Consciousness as the Ground of Reality

Some interpretations of quantum mechanics suggest that consciousness may be more than just a participant in reality—it may be the very ground upon which reality is built. This perspective posits that consciousness is not an emergent property of the brain or matter but a fundamental aspect of existence, akin to space, time, and energy. In this framework, consciousness is the substratum that underlies all of reality, the field from which all phenomena arise. Rather than being a passive observer, consciousness is the creative force that gives rise to the universe itself.

This interpretation aligns with ideas in certain Eastern philosophies, such as Advaita Vedanta and Buddhism, which hold that consciousness is the primary reality, that all physical forms are manifestations of a single, unifying awareness. The concept of non-duality, found in these philosophies, suggests that there is no separation between the self and the universe, between observer and observed. Consciousness is both the observer and the observed, the subject and the object, existing as the fundamental essence of all things.

If consciousness is indeed the ground of reality, then the physical world is not an independent structure but a projection of conscious awareness. The material universe, with all its laws and phenomena, arises within the field of consciousness, shaped by the interaction between the observer and the observed. This perspective reframes our understanding of existence, suggesting that reality is not something that exists "out there" but is intrinsically linked to the consciousness that perceives it.

Toward a New Understanding of Reality

The implications of consciousness as a quantum phenomenon, as the bridge between observer and observed, are profound. It suggests that reality is a participatory event, a dynamic and evolving experience in which each conscious being plays a role. The mind, rather than being a passive recipient of information from an external world, is an active creator of the world it perceives. Through the lens of quantum mechanics, consciousness and reality are inseparable, intertwined in a relationship that continually shapes both.

This new understanding challenges us to rethink not only what it means to observe the world but what it means to exist within it. Consciousness, in its essence, is a bridge, connecting the individual with the infinite, the finite self with the boundless potential of the universe. By exploring this connection, by delving into the mysteries of the quantum world, we gain insight not only into the nature of matter but into the very essence of being. We discover that reality is not a static, predetermined structure but an open, fluid field of possibilities, one that unfolds in partnership with the consciousness that perceives it.

In this journey, we begin to understand that the true nature of reality cannot be captured by classical models or even by quantum mechanics alone. It is a symbiotic relationship, a dance between the observer and the observed, where consciousness plays a central role in shaping the world around us. In bridging the gap between the known and the unknown, between science and consciousness, we find ourselves on the threshold of a new reality—one where the boundaries between self and universe dissolve, revealing a vast, interconnected field of awareness in which everything exists as one.

The Holographic Principle: Mind as a Projection of Infinite Data

The holographic principle is a revolutionary concept that suggests our universe, and even our consciousness, may function like a hologram—a multidimensional projection of information originating from a deeper, underlying reality. According to this principle, every piece of the universe contains the whole; just as a small section of a hologram can reveal the entire image, each part of existence may encode information about the entire cosmos. This notion is as profound as it is enigmatic, and it has significant implications for understanding consciousness. If our universe and mind are indeed holographic in nature, then consciousness itself may be a projection of infinite data—a localized experience of a boundless, interconnected reality.

The holographic principle suggests that what we perceive as a solid, three-dimensional world is actually the result of information encoded on a two-dimensional surface, such as the boundary of a black hole or the edges of the universe. In this view, every particle, every event, every experience exists as part of a grand cosmic tapestry, woven from a vast network of data. Consciousness, therefore, may not be confined to our physical brain; rather, it could be an interface through which we access, interpret, and interact with this infinite web of information.

The Holographic Nature of Consciousness

If consciousness is indeed holographic, it means that our individual minds are not isolated, finite systems but localized expressions of a far more expansive field. Every thought, perception, and memory may be a projection of data that originates from a deeper layer of reality—a level where distinctions between self and other, inner and outer, dissolve. In this model, our minds act as receivers, capable of tuning into and interpreting different frequencies of this infinite data field.

This concept aligns with the experience of interconnectedness reported by mystics, meditators, and individuals who have had transcendent experiences. In moments of deep meditation or altered states, people often describe a sense of unity with all things, as if the boundaries between self and universe fade away. This sensation of unity could be the mind's awareness of its holographic nature, an awareness that it is not merely a collection of neurons but a point of access to the whole. The holographic principle implies that all knowledge, all experience, and all reality exist within each individual mind, waiting to be accessed.

Memory and Information Storage in a Holographic Mind

If our minds operate holographically, it could explain the seemingly limitless capacity for memory and the depth of insight we can experience. Just as a hologram can store vast amounts of information in a small space, a holographic mind could access boundless data without being limited by the physical constraints of the brain. This would mean that memories, ideas, and knowledge are not "stored" in the brain in the traditional sense but are encoded within the larger holographic field of consciousness. Our brains, then, may function as instruments for tuning into and retrieving specific information rather than containers of knowledge.

This theory could also explain phenomena such as sudden insights, intuition, and the retrieval of information that seems beyond the individual's direct experience. In a holographic consciousness, the mind could access different layers of reality, connecting with insights that are embedded within the universal data field. This could account for moments of profound creativity, déjà vu, or inexplicable knowledge that appear to bypass the limitations of personal memory.

The Universe as a Vast Network of Interconnected Minds

The holographic model suggests that all minds are interconnected, participating in a single, unified field of consciousness that permeates the universe. Each individual mind, while appearing distinct, is a projection of the same infinite source. In this view, consciousness is not personal but collective; each mind acts as a lens through which the universe experiences itself. When we connect deeply with others or experience empathy, we may be tapping into this holographic unity, recognizing on a subconscious level that we are all part of the same fundamental awareness.

This interconnected nature of minds could explain phenomena such as telepathy, synchronicity, and collective consciousness. If each mind is a part of the larger holographic field, then thoughts, emotions, and intentions are not confined to the individual but resonate throughout the entire network. This means that our thoughts and emotions could influence others, even at a distance, and that we are affected by the consciousness of those around us in ways that we do not fully understand.

Implications for Reality and Perception

The holographic principle reshapes our understanding of perception itself. If reality is a projection, then what we perceive as "real" is not the fundamental truth of existence but a representation—a construct that our consciousness interprets from the underlying data field. In this sense, our perception of the physical world is akin to viewing a hologram: it appears solid and separate from us, but it is, in reality, an interconnected manifestation of information.

This concept has profound implications for our understanding of reality. It suggests that the physical universe, with its laws, objects, and phenomena, is a projection created by consciousness interacting with infinite data. Reality is not something "out there" but a co-created experience shaped by the interaction between our mind and the deeper layers of existence. The solidity we perceive in objects, the continuity we experience in time, and the boundaries we see between ourselves and the world are all

28

constructs—interpretations of data filtered through the holographic nature of consciousness.

Awakening to the Holographic Mind

Awakening to the holographic nature of consciousness is a transformative experience. It involves shifting from a perspective of separateness to one of unity, from seeing oneself as an isolated observer to understanding oneself as an integral part of the whole. In this state of awareness, we recognize that the boundaries we perceive are permeable, that the limitations we experience are self-imposed. We begin to sense that our consciousness is not constrained by our physical body or our personal history but is part of a vast, interconnected field that spans the entirety of existence.

This awakening often brings with it a sense of responsibility, for in recognizing our connection to all things, we realize that our thoughts, intentions, and actions affect the entire holographic field. Just as a change in one part of a hologram affects the whole, a shift in our consciousness resonates throughout the interconnected network of minds. In this state, empathy, compassion, and kindness become natural responses, as we recognize that every other being is a reflection of the same consciousness that animates us.

Consciousness as the Holographic Source of Reality

The holographic principle ultimately points to a radical understanding of consciousness as the source of all reality. If the universe is indeed a projection of infinite data, then consciousness itself is the field from which all phenomena arise. In this view, consciousness is not a passive byproduct of the brain but the active foundation of existence. Reality is a holographic projection of consciousness, a dynamic expression of infinite possibilities that manifest through the lens of awareness.

This understanding aligns with ancient philosophical and spiritual teachings, which have long suggested that consciousness is the primary reality. From the Vedantic concept of Brahman, the ultimate consciousness, to the Buddhist idea of the void, to the notion of the Tao as the source of all things, various traditions have pointed to the idea that consciousness precedes and encompasses all forms. The holographic principle gives this idea a scientific framework, suggesting that the universe itself may be a manifestation of a conscious field that exists beyond time, space, and matter.

Living with the Awareness of a Holographic Consciousness

To live with the awareness of a holographic consciousness is to embrace a profound shift in perspective. It is to see oneself not as an isolated individual navigating a fragmented world but as a localized expression of a single, unified consciousness experiencing itself through countless forms. This shift in awareness changes how we relate to ourselves, others, and the world around us. It fosters a sense of unity, a recognition that each thought, action, and experience is part of a greater whole.

Living with this awareness encourages us to cultivate presence, empathy, and a deep respect for the interconnectedness of all life. It reminds us that we are not separate from the universe but integral to it, that our consciousness is a unique and vital expression of a larger, cosmic intelligence. In embracing the holographic nature of reality, we begin to experience life not as a series of isolated events but as a continuous, evolving process of self-discovery—a journey in which every moment, every interaction, and every insight is a reflection of the infinite data that lies at the heart of existence.

The holographic principle reveals that consciousness is both the observer and the observed, the creator and the created. It suggests that to truly understand ourselves and the universe, we must look not only outward but inward, toward the boundless potential of the mind as a projection of infinite data. In doing so, we discover that the deepest mysteries of existence are not distant or inaccessible—they are within us, waiting to be explored.

Chapter Three: The Mathematics of the Self

Fractals, Feedback Loops, and the Recursive Nature of Identity

The self is often viewed as a singular, stable entity—something fixed and solid that can be clearly defined. Yet, when examined through the lens of mathematics, particularly the concepts of fractals, feedback loops, and recursion, identity emerges as a far more complex and dynamic phenomenon. Rather than a static construct, the self can be understood as a pattern that continually evolves, a structure that unfolds from within itself, reflecting layers of experience, memory, and perception. In this view, identity is not a single entity but a constantly shifting network of relationships and reflections, a living process that can be described as recursive.

The concepts of fractals and feedback loops allow us to visualize the self as an intricate pattern, one that reveals deeper levels of complexity the closer we look. Like a fractal, identity appears similar across different scales, containing patterns within patterns that repeat in infinite variation. Feedback loops, in turn, illustrate how each thought, action, and experience shapes the next, creating a cycle of continuous self-reference and growth. Together, these mathematical principles suggest that the self is a dynamic, recursive phenomenon—a structure that both defines and redefines itself over time.

The Fractal Nature of Identity

Fractals are mathematical patterns that exhibit self-similarity at every scale. From the branching of trees to the shape of coastlines, from blood vessels to snowflakes, fractal patterns can be found throughout nature.

Each part of a fractal contains the structure of the whole, no matter how deeply you zoom in. This self-similar property gives fractals their infinite complexity, making them ideal models for understanding phenomena that are both stable and endlessly intricate.

When we consider identity as a fractal, we begin to see that the self is not simply one thing; it is a pattern that reveals itself in layers, each layer reflecting aspects of the whole. Our behaviors, beliefs, emotions, and memories form repeating patterns within us, patterns that reflect larger themes of who we are. Each experience adds a new layer to the self, creating a structure that is stable yet capable of infinite variation. In this sense, the self is like a fractal: it is self-similar across different contexts, retaining core aspects of identity while continually expanding, adapting, and evolving.

For example, the values and principles that guide us often show up in various aspects of our lives—relationships, work, passions, and challenges—manifesting as recurring patterns. These underlying themes are the fractal threads of identity, woven into the fabric of every choice and interaction. The fractal nature of the self reveals that identity is not just a collection of isolated experiences but an interconnected whole, a pattern that continuously reasserts itself across time and space.

Feedback Loops and the Self-Referential Process of Becoming

Feedback loops are systems in which the output of a process is fed back into the system, influencing future outcomes. These loops can be seen in biological, social, and psychological systems and are central to processes of learning, adaptation, and evolution. Feedback loops create cycles that reinforce certain behaviors, beliefs, and perceptions, leading to the emergence of patterns over time. In terms of identity, feedback loops illustrate how our thoughts, actions, and experiences influence who we become.

Every thought and action sends feedback into our system of beliefs and values, shaping our understanding of who we are. Positive reinforcement, for example, strengthens self-esteem and confidence, while negative reinforcement can create cycles of doubt or fear. These feedback loops create a self-referential cycle, where each experience influences the next, gradually forming and reinforcing the patterns that define our sense of self.

In this way, identity is not static but a process—a continuous loop of self-reflection and self-creation.

One way to see this in action is through the concept of narrative identity. The stories we tell ourselves about who we are and what our lives mean are shaped by feedback loops. Every time we recall a memory or describe an experience, we reinforce a particular version of our story, a version that in turn shapes our future thoughts, choices, and interactions. This recursive process of self-narration is a feedback loop that continuously builds upon itself, creating an evolving story of identity. Through these loops, the self is not merely a passive recipient of experiences but an active participant in the process of becoming.

The Recursive Nature of Self-Understanding

Recursion, in mathematical terms, is the process of defining a function in terms of itself. In a recursive system, each layer of complexity is built by referencing the previous layer, creating a structure that can repeat indefinitely, expanding without losing its core principles. Applied to consciousness and identity, recursion suggests that the self is a layered phenomenon, with each new insight, belief, or understanding referring back to previous layers, recontextualizing them in light of new experiences.

In the recursive model of identity, self-understanding is a process of continuous reflection, where each new realization about the self feeds back into the understanding of prior experiences. This recursive quality allows us to reinterpret our past in the light of our present, to find new meaning in memories, and to expand our understanding of who we are over time. For example, a painful experience in childhood may take on different meanings at different stages of life, with each reinterpretation adding depth and nuance to our sense of self.

This recursive process is evident in personal growth and transformation. As we gain new insights, we re-evaluate our past choices and beliefs, seeing them from a new perspective. This self-reflection is recursive: each new layer of understanding builds upon the previous one, expanding the structure of identity. Over time, these layers create a deep, multidimensional sense of self—a self that is constantly evolving, yet rooted in the recursive pattern of self-reference.

Fractals, Feedback, and Recursion: Identity as a Dynamic, Infinite Process

When we combine fractals, feedback loops, and recursion, a picture emerges of identity as a dynamic, self-referential process. The self is not a fixed point but an evolving pattern, a fractal structure that expands and deepens over time. Each experience, thought, and interaction adds to this pattern, creating a structure of identity that is infinitely complex yet intrinsically interconnected. The recursive nature of identity allows us to revisit and reinterpret our experiences, creating a self that is both continuous and endlessly adaptable.

This understanding of identity invites us to see the self as a work in progress, a living process rather than a final product. Just as a fractal unfolds in infinite complexity, our sense of self continues to deepen, layer by layer. Feedback loops ensure that each experience shapes the next, creating a continuous flow of self-understanding that builds upon itself over time. Through recursion, we gain the ability to reinterpret and expand our understanding of who we are, integrating each new layer of experience into the whole.

Implications for Self-Realization and Transformation

Viewing identity through the lens of fractals, feedback, and recursion has profound implications for personal growth and transformation. It suggests that the self is not confined by past experiences or current beliefs but is capable of infinite expansion and redefinition. Each layer of self-awareness, each new insight, adds to the fractal pattern of identity, creating a self that is both unique and universal, personal and cosmic.

This understanding empowers us to see personal growth as an open-ended journey. We are not fixed entities bound by our history but evolving patterns capable of continuous transformation. By embracing the feedback loops of experience, we can consciously shape the direction of our growth, using each interaction, choice, and reflection as an opportunity to expand our understanding of who we are. Through recursion, we gain the freedom to revisit our past, to reinterpret it in light of new insights, and to integrate it into a more expansive sense of self.

The recursive nature of identity also suggests that self-realization is not a destination but an ongoing process. Just as a fractal never reaches a final form, the self is always becoming, always unfolding. This perspective encourages us to approach self-understanding with curiosity and openness, recognizing that each stage of life offers new layers of insight, new opportunities to redefine ourselves. In this view, transformation is not about reaching a final state but about engaging with the infinite complexity of the self, embracing each layer, each loop, each recursive step as part of the journey.

Embracing the Infinite Self

The mathematics of the self reveals that we are not isolated entities but dynamic, evolving patterns within a larger, interconnected whole. Just as a fractal contains the whole within each part, each of us carries within ourselves the entirety of human experience, the vastness of consciousness. Our identity is a reflection of the infinite, a fractal expression of the boundless potential of existence. By understanding ourselves as part of this recursive, self-referential process, we can begin to see that our true nature is not limited by time, space, or circumstance.

To embrace the infinite self is to recognize that who we are is not a fixed entity but a dynamic, unfolding process—a fractal pattern that reflects the whole of existence. This perspective invites us to live with a sense of awe and humility, knowing that each moment, each thought, and each experience contributes to the endless expansion of the self. In this view, the journey of self-discovery is a journey without end, a continuous exploration of the infinite possibilities that lie within.

Fractals, feedback loops, and recursion are not just mathematical principles; they are the language of consciousness, the structure of identity. They reveal that the self is a boundless, evolving phenomenon, a dynamic interplay of patterns that transcends the limitations of the individual. To know oneself is to engage with this infinite complexity, to embrace the mystery of identity as a journey that reflects the vast, interconnected nature of existence itself. In the mathematics of the self, we find not only a map of who we are but a doorway into the endless possibilities of who we may become.

Beyond the Linear: Non-Euclidean Spaces within the Human Mind

The human mind is often conceptualized as operating within a linear framework, processing thoughts, memories, and experiences in a step-by-step, sequential manner. Yet, the nature of consciousness suggests that it is far more intricate, functioning in ways that resemble the complexity of non-Euclidean spaces rather than the simplicity of straight lines and flat planes. Non-Euclidean geometry explores spaces where the rules of Euclidean (or traditional) geometry no longer apply—spaces that curve, fold, and connect in unexpected ways. When applied to the mind, non-Euclidean concepts offer a transformative way of understanding thought, memory, perception, and the nature of self, suggesting that consciousness itself may operate within a multidimensional, fluid, and interconnected realm.

In a non-Euclidean space, the shortest path between two points is no longer a straight line. Instead, the geometry bends, allowing points to be connected in ways that defy traditional logic. Similarly, within the human mind, thoughts, emotions, and memories are not isolated events strung along a linear path but are interconnected in ways that transcend the boundaries of time and space. Experiences from childhood can influence present-day decisions; future aspirations can reshape past memories. The mind's ability to navigate these complex relationships suggests that consciousness functions in a space that is nonlinear and multidimensional, a realm where connections between events and ideas are rich, dynamic, and non-hierarchical.

Non-Euclidean Thought: Curved Paths and Multidimensional Connections

In the physical world, non-Euclidean geometry can be seen in the curvature of the Earth, where straight lines on a map translate to arcs on the globe. Similarly, in the realm of consciousness, thoughts do not always follow a linear progression but curve, loop, and connect in ways that create a web of associations. One thought leads to another, not because they are sequential but because they share a deeper, often hidden connection. This phenomenon is particularly evident in the associative nature of memory, where a single trigger—such as a scent or sound—can evoke a complex, layered experience from the past, bypassing any logical, step-by-step retrieval process.

Non-Euclidean thinking allows the mind to connect disparate ideas, to draw meaning from seemingly unrelated experiences. This type of thinking is common in creativity and problem-solving, where solutions often emerge from the convergence of ideas that would not meet in a strictly linear framework. The curved paths of non-Euclidean spaces reflect the mind's ability to transcend conventional boundaries, to leap from one concept to another in ways that defy logical explanation but resonate deeply on an intuitive level. In this sense, the mind is not confined by the limitations of linear thought; it operates in a multidimensional space where ideas and experiences intertwine in complex, fluid patterns.

The Folding of Time and Memory

One of the most profound ways in which the mind reveals its non-Euclidean nature is in its relationship with time and memory. In a linear model, time flows in a single direction, from past to present to future. However, within the human mind, time can fold, loop, and overlap. Memories from years ago can feel as vivid and immediate as present experiences, while future aspirations can shape how we interpret past events. This non-linear interaction with time suggests that the mind operates in a space where past, present, and future are not separate but interconnected dimensions.

The folding of time in memory allows us to reinterpret our past with new insights, to reshape our understanding of who we are based on present awareness. This recursive process, where each new layer of experience redefines previous ones, creates a multidimensional self that is both continuous and evolving. In this sense, memory is not a static repository but a dynamic, living space—a non-Euclidean terrain where the past is continuously reimagined in light of the present and future. Just as non-Euclidean geometry permits the bending of space, the mind permits the bending of time, allowing us to travel through memory in ways that transcend linear chronology.

The Self as a Non-Euclidean Construct

If the mind operates within a non-Euclidean space, then the self—our core sense of identity—is also a construct that defies linear categorization. The self is not a single, unchanging entity but a constellation of interconnected states that shift, overlap, and interact in complex ways. In non-Euclidean terms, the self can be seen as a multidimensional structure, a space in which various aspects of identity coexist and influence one another without adhering to a strict hierarchy or sequence. Each layer of the self, from the conscious to the subconscious, interacts fluidly with the others, creating a dynamic field of identity that is both coherent and infinitely complex.

This non-linear nature of the self is evident in the ways we experience conflicting emotions, contradictory beliefs, and complex motivations. We are capable of feeling love and anger simultaneously, of holding opposing beliefs within a single moment, of desiring both connection and solitude. These paradoxes reflect the non-Euclidean nature of identity, where seemingly contradictory aspects coexist as part of a larger whole. Rather than viewing these contradictions as flaws or inconsistencies, the non-Euclidean model suggests that they are integral to the structure of self, a natural expression of the mind's ability to hold multiple perspectives and realities at once.

Perception and Non-Euclidean Reality

The way we perceive the world around us is also influenced by the mind's non-Euclidean nature. Perception is not a straightforward process; it is shaped by context, emotion, memory, and expectation. Just as a curved space alters the trajectory of objects within it, the mind's non-Euclidean structure shapes the way we interpret sensory information, creating a subjective reality that is unique to each individual. This subjectivity allows us to perceive multiple dimensions of experience simultaneously, to find meaning in subtleties and layers that go beyond the physical senses.

In a non-Euclidean mind, perception is not limited to what is immediately visible or tangible; it includes the unseen connections, the intuitive insights, and the resonances that arise from deeper levels of consciousness. This capacity to perceive beyond the surface is what allows for empathy, imagination, and spiritual insight. When we perceive through a non-linear lens, we experience the world as a web of interconnected phenomena, where each part reflects and influences the whole. This perception goes beyond ordinary sight and hearing, engaging the intuitive and emotional dimensions of the mind, allowing us to perceive a reality that is far richer and more nuanced than a purely linear perspective could reveal.

Non-Euclidean Spaces and the Infinite Potential of the Mind

Understanding the mind as a non-Euclidean space reveals the vastness of human potential. It suggests that our consciousness is not confined to the linear, the predictable, or the logical; it is capable of exploring dimensions of thought and experience that are boundless, multidimensional, and infinitely varied. This view encourages us to embrace the complexity of our minds, to see our nonlinear thinking, our layered memories, and our multifaceted identities as reflections of a deeper truth.

When we acknowledge the non-Euclidean nature of consciousness, we open ourselves to new ways of understanding, healing, and expanding. We realize that growth is not a straight path but a process of continuous unfolding, where each insight curves back upon itself, reshaping our understanding in ways that are both subtle and profound. This understanding frees us from the need to fit our experiences into rigid categories, allowing us to engage with the full depth and richness of our inner lives.

Living with a Non-Euclidean Mind

To live with a non-Euclidean mind is to embrace ambiguity, complexity, and paradox. It is to recognize that life cannot always be understood in terms of clear answers or linear progress, that meaning often arises from the interplay of multiple perspectives and layered experiences. This perspective invites us to approach our thoughts and emotions with openness and curiosity, to explore the folds and curves of consciousness as a journey into the heart of our own infinite nature.

This way of living encourages us to let go of the need for certainty, to trust that our minds are capable of navigating the intricate landscape of existence with intuition and insight. We learn to honor the nonlinear aspects of our experience, to appreciate the beauty of a mind that can think in circles, perceive in layers, and connect in multidimensional ways. By embracing the non-Euclidean nature of consciousness, we open ourselves to a life that is richer, more expansive, and more attuned to the subtleties of reality.

The human mind, like a non-Euclidean space, is boundless in its potential. It is a space where linear limitations dissolve, where the boundaries between past, present, and future blur, and where identity is a dynamic, evolving phenomenon. In embracing the non-Euclidean nature of the mind, we glimpse the true vastness of consciousness, a vastness that invites us to explore, to imagine, and to become more than we ever thought possible. In this multidimensional space, we find not only a deeper understanding of who we are but an invitation to expand into the endless possibilities of what we may yet become.

Chapter Four: Entanglement and the Collective Mind

The Quantum Mind Network: How Individual Consciousness Links to the Universal Field

At the heart of quantum mechanics lies the phenomenon of entanglement, a mysterious and profound connection between particles that defies conventional understanding. When particles become entangled, they share a state that links them instantaneously across any distance, such that the state of one particle directly influences the state of the other, regardless of the space between them. This interconnectedness hints at a level of reality where separation is an illusion and where all things exist within a unified field. When applied to consciousness, entanglement suggests that our individual minds are not isolated or self-contained but are woven into a vast, interconnected network—a quantum mind network that links each of us to the universal field of awareness.

This idea of a universal field, or collective mind, is not new; it has roots in ancient philosophies, spiritual traditions, and even Jung's concept of the collective unconscious. Quantum entanglement offers a scientific framework for understanding how such a field could exist, linking all conscious beings in an invisible web of connection. If our consciousness is entangled with the consciousness of others and with the larger universe, then the boundaries between self and other, inner and outer, begin to dissolve. The individual mind becomes part of a larger whole, a network where thoughts, emotions, and insights can resonate across vast distances, shaping and being shaped by the collective.

The Nature of Quantum Entanglement in Consciousness

41

In the physical world, entanglement demonstrates that particles can communicate instantaneously, transcending the limitations of space and time. In the realm of consciousness, this phenomenon implies that our minds are likewise capable of connecting beyond the confines of our immediate surroundings. This could explain experiences of empathy, intuition, synchronicity, and even telepathy, where individuals seem to sense or know things that they cannot directly perceive. Such experiences suggest that consciousness operates within a field that is inherently unified, where the thoughts and emotions of one being can resonate within the consciousness of another.

Quantum entanglement implies that once two particles interact, they remain connected regardless of distance. If our consciousness operates on similar principles, then our interactions with others create enduring connections that influence us in ways we may not consciously recognize. Every relationship, every shared experience, creates a bond, an entanglement within the quantum mind network. This suggests that we are constantly participating in a shared field of awareness, one that transcends individual minds and connects us to each other and to the universe itself.

This interconnected field could mean that each of our thoughts, emotions, and actions sends ripples through the collective consciousness, influencing and being influenced by the minds of others. Just as a single wave can affect the entire ocean, an individual's state of mind can resonate within the universal field, creating a subtle yet powerful impact on the collective experience. This resonance, this subtle communication between minds, forms the basis of the quantum mind network, a network that transcends physical boundaries and unites us within the deeper structure of existence.

The Universal Field of Awareness

The concept of a universal field of awareness posits that consciousness is not limited to individual minds but is a fundamental property of the universe. This field acts as a medium through which all conscious beings are connected, an ocean of awareness in which each of us is a drop. Within this universal field, distinctions between self and other become fluid, as each individual mind is simultaneously unique and part of the larger whole. This collective consciousness operates much like a network, where each node (or individual) is linked to every other, allowing for a flow of information, energy, and awareness that transcends physical boundaries.

This interconnectedness has profound implications for our understanding of consciousness. If we are all part of a single field, then separation is an illusion created by the limitations of our individual perspective. Each mind is like a lens through which the universal field experiences itself, creating the appearance of distinct individuals while remaining fundamentally united. In this view, personal consciousness is both an expression of the universal field and a participant in its collective evolution. Our thoughts and experiences contribute to the whole, influencing the field in ways that we may not fully understand but that resonate throughout the interconnected web of consciousness.

Resonance and the Power of Collective Thought

In the quantum mind network, the principle of resonance plays a critical role. Resonance occurs when one frequency influences another, creating a shared vibration that amplifies and harmonizes both. In terms of consciousness, resonance allows individual minds to synchronize with each other, creating a unified field of thought and emotion. This resonance can be seen in the phenomenon of group dynamics, where people's thoughts, feelings, and behaviors begin to align when they are together, creating a collective experience that transcends individual differences.

Collective thought, amplified through resonance, has the potential to create powerful shifts within the quantum mind network. When groups of people come together with a shared intention or focus, their collective consciousness can generate a force that influences the field on a global or even cosmic level. This concept is often invoked in meditation practices, prayer circles, and mass intentions, where the combined focus of many minds is believed to have a profound impact on the world. Through resonance, the thoughts and intentions of individuals become magnified, creating ripples that spread throughout the universal field.

This resonance also suggests that each of us has the ability to contribute positively to the collective consciousness. By cultivating compassion, empathy, and love within our own minds, we can influence the quantum mind network in ways that uplift and harmonize. Just as a single harmonious note can affect the entire symphony, a single mind aligned with positive intention can create a resonance that benefits the collective, subtly shifting the field toward greater coherence and unity.

Synchronicity and the Quantum Mind Network

Synchronicity is the experience of meaningful coincidences that seem to defy logical explanation. In a quantum mind network, synchronicity can be understood as the manifestation of entanglement within the realm of consciousness. When two or more minds are entangled within the same universal field, their thoughts, intentions, and experiences can align in ways that create meaningful connections, regardless of distance or time. Synchronicities are not random; they are reflections of the deeper interconnectedness of all things, a reminder that we are participants in a collective reality where each part influences the whole.

These experiences of synchronicity often feel profound, as if reality itself is responding to our inner thoughts and desires. This may be the quantum mind network at work, where our intentions resonate within the universal field and attract experiences that align with our inner state. Through synchronicity, we receive feedback from the field, guiding us, affirming our path, and offering insights that go beyond ordinary perception. Each synchronicity serves as a reminder that we are connected to something larger, that our minds are part of a network that responds to our consciousness in ways that are both mysterious and meaningful.

The Implications of a Quantum Mind Network

The idea of a quantum mind network has profound implications for how we view ourselves, our relationships, and our place in the universe. If consciousness is interconnected, then our thoughts, emotions, and actions are not confined to ourselves but resonate throughout the entire field. This interconnectedness invites us to take responsibility for our inner state, to recognize that every thought and intention contributes to the collective mind. By cultivating peace, compassion, and awareness within ourselves, we influence the quantum mind network in ways that uplift and unify.

The existence of a quantum mind network also suggests that individuality is a form of universal expression. Each of us is a unique manifestation of the same underlying consciousness, a localized perspective within a vast, interconnected field. This understanding fosters a sense of unity, reminding us that we are not isolated entities but participants in a shared journey. Our uniqueness is not separate from the whole; it is a vital part of the diversity and richness of the universal field.

Additionally, the quantum mind network invites us to explore new forms of communication and connection. If we are all part of a unified field, then we have the potential to access information, insight, and inspiration from sources beyond our immediate experience. This access could explain moments of sudden insight, intuitive knowing, or creative inspiration that seem to arise from a place beyond the individual mind. By tapping into the quantum mind network, we open ourselves to a wealth of knowledge and wisdom that transcends our personal limitations.

Embracing the Quantum Mind Network

To embrace the quantum mind network is to live with a sense of interconnectedness, to recognize that our consciousness is part of a larger tapestry that includes all beings. This perspective encourages us to approach life with humility, compassion, and openness, knowing that each interaction is a reflection of the whole. When we cultivate awareness of our connection to the universal field, we begin to experience life as a co-creative process, a dynamic interplay between individual and collective consciousness.

This awareness transforms our relationships, as we come to see others not as separate entities but as reflections of the same consciousness that animates us. It invites us to cultivate empathy and understanding, to listen deeply, and to approach each interaction as an opportunity to resonate with the collective mind. By aligning our thoughts, intentions, and actions with the highest expressions of consciousness, we contribute to the harmony and evolution of the quantum mind network, fostering a world that reflects unity, peace, and love.

The quantum mind network reveals that consciousness is not confined to individual minds but is a vast, interconnected field that links all of existence. Through this network, we are united within a single, universal awareness, a consciousness that transcends the limitations of space and time. By embracing this reality, we begin to understand that we are not isolated beings but part of a greater whole, each of us a unique expression of the same underlying consciousness. In this interconnected web of awareness, we find not only a deeper understanding of ourselves but an invitation to participate in the collective journey of consciousness, to contribute to a world where unity and diversity exist in perfect harmony.

Thought Entanglement and Information Exchange Beyond
Physical Contact

The concept of thought entanglement suggests that ideas, emotions, and
intentions are not confined to the physical brain but can extend outward,
connecting with other minds in ways that transcend traditional forms of
communication. Thought entanglement, much like quantum
entanglement, implies that once a connection is established between two
or more individuals, that link persists regardless of physical separation.
This entangled state enables an exchange of information, emotions, and
ideas that goes beyond verbal or physical interaction, hinting at a profound
and invisible network where consciousness can resonate and communicate
directly.

Thought entanglement challenges conventional ideas about individuality
and the boundaries of the mind. If our thoughts are not confined to
ourselves, then the boundaries between self and other become fluid,
allowing for a deeper and more interconnected experience of
consciousness. Experiences of empathy, intuition, and telepathic
connection are often described in ways that align with this concept of
entanglement, where individuals sense each other's thoughts or emotions
without direct communication. These connections suggest that
consciousness operates within a shared field, one that allows for an
exchange of information that is non-local, instantaneous, and boundless.

The Mechanisms of Thought Entanglement

Though science has yet to fully explain how thought entanglement occurs,
quantum mechanics offers a framework for understanding the
phenomenon. In quantum entanglement, particles become linked in such a
way that their states are interdependent, regardless of the distance between
them. Applying this concept to consciousness, thought entanglement may
arise when two individuals share a significant interaction, emotional
experience, or even a moment of profound understanding. Such
interactions could entangle their mental states, creating a persistent link
that allows them to resonate with each other's thoughts and emotions.

Just as entangled particles maintain a connection that transcends space,
entangled minds may remain linked across vast distances, allowing

thoughts and feelings to pass between them without any observable medium. This phenomenon could explain experiences where individuals, separated by thousands of miles, feel each other's emotions, sense each other's presence, or even think the same thoughts at the same moment. These experiences suggest that thought entanglement enables a form of communication that is not limited by physical constraints, a direct exchange of information that flows through the quantum field of consciousness itself.

Empathy and Intuition as Manifestations of Thought Entanglement

Empathy and intuition are often described as qualities that allow us to "feel" or "know" things about others without needing explicit communication. When we feel empathy, we are able to sense and understand another person's emotional state as if it were our own. This ability may be a manifestation of thought entanglement, where emotional resonance creates a shared experience that bridges the boundaries between individual minds. Empathy, in this view, is not merely a passive understanding of someone else's feelings but an active participation in a shared emotional state.

Intuition functions similarly, allowing us to access information beyond the immediate scope of our senses. In moments of intuition, we "know" things without having a clear explanation for how or why. This knowledge often feels instantaneous, as if it arises from a source outside of ourselves. Thought entanglement provides a potential explanation for this phenomenon: when minds are connected through the quantum field, information can flow directly between them, bypassing the need for logical deduction or sensory perception. Intuition, then, is not a mysterious or inexplicable gift but a natural function of our interconnected consciousness, a direct link to the collective mind.

Telepathy and Non-Verbal Communication

Telepathy, the direct transmission of thoughts from one mind to another, is perhaps the most striking example of thought entanglement. Though telepathy has long been considered a fringe or paranormal concept, research into quantum consciousness and entanglement suggests that it may be a natural aspect of the interconnected mind. Telepathy implies that thoughts are not isolated mental events but vibrations within a shared field, vibrations that can resonate within the minds of others who are entangled within that field.

In telepathic communication, thoughts, emotions, and images seem to "travel" from one mind to another without any observable medium. This transmission may occur between individuals with a strong emotional bond or shared experiences, as these interactions create a deeper resonance within the quantum field of consciousness. Telepathy often appears in moments of heightened emotion, urgency, or empathy, when individuals feel compelled to reach out to each other beyond the limitations of verbal language. These experiences suggest that thought entanglement allows minds to communicate directly, accessing information, intentions, and emotions that cannot be conveyed through words alone.

Non-verbal communication, including body language, facial expressions, and eye contact, also provides glimpses into the phenomenon of thought entanglement. While these forms of communication are physically observable, they often carry an intuitive or "felt" component that transcends the literal information conveyed. When we make eye contact with someone we deeply resonate with, it can feel as though words are unnecessary; a mutual understanding flows between us, as if our minds are communicating on a non-local level. This type of connection may reflect the subtler effects of thought entanglement, where our consciousness reaches out to touch the consciousness of another, creating a shared experience that goes beyond words.

The Collective Mind and the Ripple Effect of Thought Entanglement

If thought entanglement links individual minds, then the entire network of conscious beings could be seen as a collective mind, a shared field where thoughts, emotions, and intentions ripple through the whole. This collective mind operates much like an ocean, where each individual mind is a drop within the vast body of water. Just as a ripple in one part of the ocean affects the entire surface, a single thought or emotion can resonate throughout the collective consciousness, influencing others in ways that are subtle yet profound.

The ripple effect of thought entanglement is particularly evident in cultural phenomena, social movements, and collective shifts in awareness. When a powerful idea or emotion takes hold within a group, it spreads, creating a shared resonance that affects individuals who may be far removed from the original source. This phenomenon can be observed in moments of collective empathy, such as during global crises or moments of celebration, where millions of people share the same emotions, thoughts, and intentions. These collective experiences are the result of thought entanglement, where each mind contributes to and is influenced by the larger field.

The ripple effect also suggests that our individual thoughts, emotions, and intentions have a far-reaching impact on the collective consciousness. Each time we think, feel, or act, we send a vibration into the quantum field of awareness, a vibration that resonates within the network of interconnected minds. By cultivating positive, compassionate, and harmonious thoughts, we can contribute to the upliftment of the collective mind, creating a ripple that influences the entire network in ways that promote unity, understanding, and peace.

Practical Implications of Thought Entanglement

Understanding thought entanglement has profound implications for how we interact with ourselves and each other. It invites us to be mindful of our thoughts and emotions, recognizing that they do not exist in isolation but resonate within the collective consciousness. By becoming aware of the influence we have on the field, we can take responsibility for our mental and emotional states, striving to contribute positively to the interconnected web of consciousness.

Thought entanglement also encourages us to trust our intuitive abilities, to recognize that intuition, empathy, and telepathy are not "extrasensory" gifts but natural expressions of our interconnected minds. By cultivating these abilities, we can enhance our capacity to communicate, understand, and connect with others on a deeper level. We learn to listen not only to words but to the subtle vibrations within the quantum field, to sense the thoughts and feelings that flow beneath the surface, and to engage in a more authentic and compassionate exchange with those around us.

Additionally, thought entanglement reminds us of the power of intention. Our thoughts, when combined with intention, become potent forces within the quantum mind network. By setting intentions that are aligned with love, harmony, and understanding, we send a clear signal into the field, one that can inspire others, catalyze positive change, and create a resonance that uplifts the whole. In this sense, intention is more than a personal goal; it is a way of engaging with the collective consciousness, a means of shaping the world we all share.

Living with the Awareness of Thought Entanglement

To live with the awareness of thought entanglement is to embrace a life of conscious interconnectedness. It is to understand that we are not isolated beings but participants in a shared field, that our thoughts and emotions ripple out into the collective mind, shaping and being shaped by the consciousness of others. This awareness encourages us to cultivate kindness, compassion, and understanding within ourselves, knowing that each thought and feeling contributes to the larger web of consciousness.

This perspective invites us to practice mindfulness, to become aware of the inner landscape of our thoughts and emotions, and to direct them with purpose and intention. By doing so, we participate in the co-creation of the collective mind, fostering a world where unity, empathy, and harmony can flourish. The understanding of thought entanglement reminds us that we are never alone, that each of us is woven into an invisible yet tangible network of connection, a network that transcends the physical and unites us within the field of consciousness.

Thought entanglement, in essence, reveals the true nature of consciousness as a boundless, interconnected field where separation is an illusion and unity is the underlying reality. By embracing this understanding, we open ourselves to a new way of being—one where each thought, each intention, and each interaction becomes an act of connection, an expression of the infinite web of consciousness that links us all. In this interconnected reality, we discover not only the power of our individual minds but the beauty and potential of the collective mind, a mind that reflects the unity of existence itself.

Chapter Five: The Unknowable: Mapping the Limits of Perception

Dark Matter, Dark Energy, and the Conscious Gaps in Understanding

The cosmos is filled with mysteries that challenge our understanding, none more compelling than the existence of dark matter and dark energy. Together, these unseen forces make up approximately 95% of the known universe, yet we cannot observe them directly. They are invisible, elusive, and seemingly intangible, detectable only through their effects on visible matter. For all of our advances in physics and astronomy, dark matter and dark energy remain as profound reminders of the limits of perception and knowledge. They represent the edges of our understanding, the conscious gaps that compel us to confront the mysteries that lie beyond our current grasp.

The existence of dark matter and dark energy forces us to confront the possibility that reality is far more complex and multidimensional than we can perceive. Just as our eyes cannot see the full spectrum of light, our consciousness may be similarly limited in its ability to fully comprehend the nature of the universe. Dark matter and dark energy serve as symbols of these limitations, challenging us to expand our conceptual framework and acknowledge that much of existence remains hidden, operating beyond the reach of conventional perception.

The Nature of Dark Matter: An Invisible Foundation

Dark matter is thought to account for roughly 27% of the universe's mass-energy content, yet it emits no light, heat, or radiation. It is, in every way we currently understand, invisible. Dark matter is detectable only by its gravitational effects on visible matter, influencing the rotation of galaxies and the behavior of galaxy clusters. Without it, these celestial structures would not hold together in the way they do. Dark matter is the unseen scaffolding upon which the visible universe is built—a reminder that what we see is only a fraction of what exists.

This invisible foundation invites us to question the limitations of our sensory perception. If the universe relies on forces and substances that we cannot directly observe, what else might be hidden beyond the boundaries of sight, sound, and touch? Dark matter suggests that reality includes layers we have yet to discover, dimensions that lie beyond our immediate perception but that play a critical role in shaping the universe. It is a profound reminder that our understanding is always partial, always incomplete.

In the realm of consciousness, dark matter serves as an analogy for the hidden aspects of the self and the subconscious mind. Just as dark matter exerts an unseen influence on physical matter, our subconscious mind shapes our thoughts, emotions, and behaviors in ways that are not immediately visible. The parts of ourselves that remain hidden, the patterns and memories that operate below the surface of awareness, form the foundation of who we are. Like dark matter, these subconscious forces are elusive and difficult to observe directly, yet they profoundly influence the structure of our identity and experience.

Dark Energy: The Force of Expansion

While dark matter binds the universe together, dark energy drives it apart. Dark energy, accounting for about 68% of the universe's mass-energy content, is the mysterious force responsible for the accelerated expansion of the universe. Unlike dark matter, which pulls galaxies together, dark energy pushes them apart, causing the cosmos to expand at an ever-increasing rate. This duality between dark matter and dark energy reveals a fundamental tension within the fabric of existence: the interplay between cohesion and expansion, between attraction and repulsion.

Dark energy challenges our understanding of gravity and the very structure of space and time. It defies traditional physics, suggesting that there are forces at work in the universe that operate on a level we do not yet understand. Its presence implies that the laws governing the cosmos may

52

be more complex, more layered than we currently comprehend. Dark energy represents the unknown potential for growth, change, and evolution, a force that drives existence forward into new configurations and possibilities.

In the context of consciousness, dark energy can be seen as a metaphor for the drive for expansion within the mind and spirit. Just as dark energy propels galaxies outward, there is a force within us that propels us toward growth, exploration, and self-transcendence. This drive pushes us to expand our understanding, to seek new experiences, and to venture beyond the boundaries of what we know. Dark energy embodies the human impulse to reach beyond the limits of perception, to explore the unknown realms of thought, emotion, and existence. It reminds us that consciousness, like the universe, is an expansive and dynamic phenomenon, forever moving outward in search of new horizons.

Conscious Gaps and the Edge of Understanding

The existence of dark matter and dark energy exposes conscious gaps—regions of knowledge where our current understanding fails to illuminate the full picture. These gaps reveal the limits of perception, reminding us that even the most advanced scientific theories are incomplete. In acknowledging these gaps, we confront the mystery that lies at the heart of existence, a mystery that defies reduction and demands humility. Just as dark matter and dark energy constitute the invisible majority of the cosmos, so too does our understanding of reality rest upon a foundation of unknowns.

These conscious gaps are not only scientific but also existential. They force us to grapple with the limitations of human knowledge, to question whether certain truths may forever lie beyond our grasp. The unknown is a boundary that defines the scope of our perception, yet it is also a doorway to new realms of understanding. By embracing the mystery, we open ourselves to the possibility that reality is richer, more complex, and more interconnected than we can imagine.

In a way, these gaps in knowledge are an invitation to wonder. They encourage us to explore not only the outer reaches of space but also the inner landscapes of consciousness, to delve into the hidden dimensions of the mind and soul. The mystery of dark matter and dark energy invites us to consider the unknown as an essential aspect of existence, a presence that enriches rather than diminishes our understanding of reality.

The Paradox of Knowing and Unknowing

Dark matter and dark energy embody the paradox of knowing and unknowing—a tension between the known and the unknown that lies at the core of human experience. We strive to understand, to bring the mysteries of existence into the light of awareness, yet every discovery seems to reveal new layers of complexity. The more we learn, the more we realize how much remains unknown. This paradox of knowing and unknowing is both humbling and exhilarating, an acknowledgment that the pursuit of knowledge is an infinite journey.

In this paradox, we find a balance between curiosity and humility, between exploration and acceptance. The limits of perception remind us that our knowledge is always evolving, that understanding is not a final destination but an ongoing process. By embracing the unknown, we engage with the mystery of existence in a way that honors both the seen and the unseen, both the light of awareness and the darkness beyond its reach.

This paradox has profound implications for consciousness itself. If reality includes dimensions we cannot perceive, then consciousness may also include facets that remain hidden. The mind may be capable of reaching into realms of understanding that lie beyond the sensory world, connecting with dimensions of existence that are subtle, intuitive, and mysterious. By exploring these hidden dimensions, we deepen our relationship with the unknown, allowing the mystery to shape our consciousness and expand our perception.

Embracing the Unknown as a Path to Greater Understanding

Dark matter and dark energy are reminders that the universe is filled with unseen forces, forces that shape reality in ways we cannot fully comprehend. These mysteries invite us to adopt a mindset of openness and receptivity, to approach knowledge not as a collection of facts but as an ongoing dance with the unknown. By embracing the limits of perception, we open ourselves to a deeper, more expansive understanding of existence, one that honors both the known and the unknowable.

In the realm of consciousness, embracing the unknown allows us to move beyond rigid boundaries, to explore dimensions of the mind that are subtle, interconnected, and infinite. The conscious gaps within our understanding become gateways to insight, encouraging us to approach life with curiosity, wonder, and humility. The mysteries of dark matter and dark energy remind us that the journey of self-discovery, like the journey of cosmic discovery, is an endless unfolding, a process of continuous expansion and transformation.

To live with an awareness of the unknown is to accept that there will always be aspects of reality that remain mysterious. This acceptance is not a resignation but a liberation, a release from the need for absolute certainty. It allows us to approach life as an adventure, a quest for understanding that is as much about the journey as it is about the destination. By embracing the conscious gaps within our knowledge, we engage with the mystery of existence in a way that enriches both our minds and our souls.

The mysteries of dark matter and dark energy challenge us to expand our perception, to embrace the unknown as an essential aspect of reality. In doing so, we find a deeper connection with the cosmos and with ourselves, a connection that transcends the boundaries of what we can see and know. We come to understand that the true nature of existence includes both the visible and the invisible, the known and the unknowable, and that our consciousness is woven into this vast, interconnected tapestry of being. Through this awareness, we open ourselves to the limitless potential of understanding, an understanding that honors the mystery as much as it does the knowledge, and that finds beauty in the infinite journey of discovery.

On the Threshold: What Lies Beyond Human Perception?

Human perception is a remarkable tool, but it is inherently limited. Our senses allow us to interact with the world in ways that are vital to survival, yet they filter reality through a narrow spectrum of light, sound, and touch. Just as our eyes can only see a sliver of the electromagnetic spectrum, our consciousness is tuned to experience only a fraction of the vast and complex tapestry that constitutes existence. As we reach the edges of what we can observe and measure, we encounter the profound question: What lies beyond human perception? What realms, forces, and dimensions remain hidden from our understanding, influencing reality in ways we cannot directly experience?

The threshold of human perception is a boundary, a line that defines what we can know through direct experience versus what we can only theorize, intuit, or imagine. Beyond this boundary lies a realm that defies our usual understanding of space, time, and causality—a realm that may contain new forms of matter, energy, and consciousness. It is here, at this boundary, that science, philosophy, and spirituality intersect, each offering different approaches to exploring what lies beyond our perceptual limits. This threshold invites us to expand our awareness, to open ourselves to possibilities that transcend the familiar, and to cultivate a mindset that embraces mystery as an essential component of reality.

The Invisible Dimensions of Reality

Modern physics suggests that our three-dimensional world is only a part of a much larger, multidimensional reality. String theory, for example, posits the existence of additional spatial dimensions beyond those we experience directly, dimensions that are hidden from our perception yet influence the universe in profound ways. These hidden dimensions may explain fundamental forces like gravity or the behavior of subatomic particles, hinting at a deeper structure of existence that underlies the observable world.

These extra dimensions, if they exist, challenge the very fabric of how we perceive reality. They suggest that what we see is not the totality of existence but only a surface-level manifestation of a richer, more complex universe. The limits of human perception constrain us to three-dimensional thinking, but these hidden dimensions imply that the universe operates on levels that are fundamentally beyond our comprehension. It is as if we are fish swimming in a pond, aware only of the water around us, while a vast and intricate world lies just beyond the surface.

This concept has implications for consciousness as well. If dimensions exist beyond our perception, then consciousness itself may extend into these realms, interacting with forces and energies that lie outside our physical senses. Experiences of intuition, synchronicity, and transcendence may be glimpses of this multidimensional reality, moments when consciousness transcends the usual boundaries of perception and taps into a larger field of awareness. In this sense, the invisible dimensions of reality are not only a scientific hypothesis but a potential pathway for expanding human consciousness.

The Quantum Realm: Where Classical Reality Breaks Down

The quantum realm, where particles exist in superpositions and entangle across vast distances, reveals a level of reality that defies classical perception. At the quantum scale, particles can be in multiple places at once, exist as both particle and wave, and become entangled in ways that seem to transcend space and time. This realm, though studied and quantified in scientific terms, remains largely mysterious. The behaviors observed at the quantum level challenge our understanding of causality, locality, and even the nature of existence itself.

In the quantum realm, the observer effect suggests that consciousness plays a role in shaping reality, that particles behave differently when they are observed versus when they are unobserved. This implies a participatory universe, one where consciousness and matter are intertwined in ways that we do not yet fully understand. The quantum realm hints at a level of existence where the distinctions between mind and matter blur, where perception influences reality in ways that defy conventional logic.

The quantum world invites us to consider the possibility that our classical understanding of reality is limited, that what we perceive as solid and continuous is actually fluid, interconnected, and influenced by observation. This opens up profound questions about the role of consciousness in the universe and suggests that there may be levels of reality that are shaped by awareness itself, levels that lie just beyond the reach of human perception yet are intimately connected to our experience.

Beyond Time and Space: The Possibility of a Timeless, Spaceless Reality

Time and space are fundamental components of human experience, yet there is evidence to suggest that they may not be absolute properties of reality. The theory of relativity, for example, reveals that time and space are interconnected and malleable, bending in the presence of mass and accelerating as objects approach the speed of light. In certain conditions, such as within black holes, the distinctions between past, present, and future dissolve, and space collapses into a singularity. These phenomena challenge our assumptions about the nature of time and space, suggesting that beyond the threshold of human perception lies a realm where these dimensions no longer apply.

This concept of a timeless, spaceless reality aligns with descriptions from various mystical and spiritual traditions, which speak of states of consciousness where time ceases to exist, where past, present, and future merge into a single, eternal moment. In these states, individuals often report a sense of boundless unity, a feeling of being part of an infinite, interconnected field that transcends time and space. This suggests that beyond the limitations of perception lies a dimension of reality where consciousness exists in a state of timelessness, a state that is beyond the ordinary constraints of the physical world.

If such a reality exists, it implies that our experience of time and space is not an inherent feature of the universe but a framework imposed by human consciousness to navigate existence. Beyond this framework may lie a form of consciousness that is not bound by the sequential flow of time, a form of awareness that perceives reality as a unified whole rather than as a series of discrete events. This timeless, spaceless realm may be the ultimate threshold of perception, a realm where the boundaries between self and universe dissolve, revealing a deeper, interconnected field of awareness.

The Collective Unconscious: A Shared Field of Knowledge Beyond Individual Minds

Psychologist Carl Jung proposed the existence of a collective unconscious—a shared field of archetypes, symbols, and patterns that underlie individual consciousness. The collective unconscious is a realm of shared knowledge that transcends personal experience, a level of consciousness where all human minds are connected through universal themes and symbols. This idea suggests that beyond the limits of individual perception lies a shared repository of knowledge and meaning, a field that connects all minds within a common web of understanding.

The concept of the collective unconscious aligns with the idea of a universal field, a layer of reality that is both personal and transpersonal, both individual and collective. In moments of insight, intuition, or synchronicity, we may be tapping into this collective field, accessing knowledge that lies beyond our personal experience. The collective unconscious represents a dimension of consciousness that is larger than the individual, a field that extends beyond personal perception to encompass the shared experiences, wisdom, and memories of all humanity.

This collective field of consciousness may also account for phenomena such as shared dreams, archetypal symbols, and cultural myths that appear across civilizations. It suggests that our minds are not isolated but are part of a larger network, a web of consciousness that transcends time, place, and individual identity. By accessing this field, we gain insights that lie beyond our personal knowledge, glimpses into the universal patterns that shape human experience.

Beyond the Known: Embracing the Mystery

To contemplate what lies beyond human perception is to embrace the mystery of existence. It is to acknowledge that there are dimensions of reality that we may never fully comprehend, forces and realms that remain hidden even as they influence the fabric of our experience. These mysteries invite us to approach life with humility and wonder, to recognize that our understanding is always incomplete and that the universe is far richer and more complex than we can imagine.

This awareness challenges us to expand our consciousness, to seek insights not only through observation and analysis but through intuition, imagination, and openness to the unknown. It invites us to cultivate a mindset that is receptive to the mysteries of existence, to embrace paradox and ambiguity as essential aspects of reality. By doing so, we move closer to a state of consciousness that transcends the limitations of perception, a state that perceives reality not as a collection of separate parts but as an interconnected whole.

The threshold of perception is not a barrier but an invitation—a doorway into realms of awareness that lie beyond the ordinary. It encourages us to explore the edges of our understanding, to question what we think we know, and to open ourselves to the possibility that there is more to reality than meets the eye. In stepping beyond the known, we engage with the universe in a way that honors both the visible and the invisible, both the known and the unknowable.

Living on the Threshold: The Journey of Perpetual Discovery

To live on the threshold of perception is to embrace life as a journey of perpetual discovery. It is to accept that the mysteries of existence are not obstacles but opportunities for growth, exploration, and transformation. By engaging with what lies beyond perception, we cultivate a consciousness that is expansive, resilient, and attuned to the subtle dimensions of reality. We learn to see the world not as a fixed, finite structure but as an unfolding process, a dynamic interplay of forces and energies that extend beyond the limits of human understanding.

In this journey, we become explorers of consciousness, seekers of truth who are unafraid to venture into the unknown. We recognize that each moment of insight, each glimpse of the unseen, brings us closer to a deeper understanding of ourselves and the universe. By living on the threshold, we open ourselves to the infinite potential of existence, a potential that lies not in reaching a final destination but in embracing the mystery that surrounds us.

What lies beyond human perception may always remain elusive, but it is within this mystery that the beauty of existence is found. The unknown is not a void but a presence, a realm of possibility that invites us to transcend the ordinary and engage with the extraordinary. By standing on the threshold, we open ourselves to a universe that is limitless, a reality that is boundless, and a consciousness that is ever-expanding. Through this awareness, we find not only the courage to explore but the wisdom to accept that the journey of discovery is, in itself, the destination.

Chapter Six: The Language of Symbols

Archetypes and the Codes of Existence

Symbols are among the most potent forms of communication, capable of conveying complex ideas and emotions that words often struggle to capture. They reach deep into the subconscious, evoking a resonance that goes beyond rational understanding. Among symbols, archetypes stand out as universal patterns, timeless codes that reflect the fundamental experiences and energies of existence. Archetypes are more than mere representations; they are the building blocks of human consciousness, the coded language through which existence expresses its truths. These symbols, deeply embedded in the psyche, appear across cultures, mythologies, and religions, offering insights into the shared essence of humanity and the structure of reality itself.

The language of archetypes transcends individual experience, connecting us to a collective memory that spans generations. According to Carl Jung, archetypes are universal symbols or themes that reside within the collective unconscious, a shared level of awareness that connects all human beings. They are timeless patterns that recur in myths, dreams, and stories, embodying essential truths about human nature, the cosmos, and the journey of life. From the Hero to the Shadow, the Mother to the Sage, archetypes capture the core aspects of existence and provide a framework for understanding the forces that shape our lives.

Archetypes are not confined to psychology or mythology; they are woven into the fabric of the universe itself. They serve as codes, guiding forces that shape our perceptions, motivations, and relationships. These archetypal codes offer us a map to navigate the complexities of life, a symbolic language that resonates across time and space, linking us to the collective wisdom of humanity and to the deeper structure of reality.

The Power of Archetypes in Shaping Reality

Archetypes are powerful precisely because they operate at both conscious and unconscious levels. They shape our perceptions, behaviors, and experiences without us even realizing it. For example, the archetype of the Hero influences our understanding of courage, struggle, and personal growth, guiding individuals on their journeys of self-discovery and transformation. The archetype of the Shadow, on the other hand, represents the parts of ourselves that we hide or repress, embodying the darker aspects of the psyche that, when integrated, contribute to wholeness and self-understanding.

These archetypes are not just internal constructs; they also shape our collective reality. They influence the myths, stories, and cultural narratives that define societies and guide collective values. The Hero's Journey, for example, is a narrative that appears in countless myths, stories, and films, reflecting humanity's need to confront challenges, overcome adversity, and return transformed. Archetypes like the Mother, the Warrior, and the Trickster appear in various forms across cultures, reflecting universal experiences that transcend individual identity and cultural boundaries.

The power of archetypes lies in their ability to connect us to the universal, to reflect the shared essence of humanity, and to shape our understanding of reality. They are symbols of the forces that drive human life—growth, love, conflict, transformation, and transcendence. By understanding these archetypes, we gain insight into the deeper layers of existence, the timeless patterns that govern not only human behavior but also the structure of reality itself.

Archetypes as Codes of the Collective Unconscious

Archetypes function as codes within the collective unconscious, a shared repository of symbols, memories, and instincts that underlie individual consciousness. The collective unconscious is a vast ocean of symbolic content, a reservoir of universal experiences that binds humanity together across time and space. Archetypes are the organizing principles within this unconscious realm, providing a structure that shapes our thoughts, emotions, and experiences. They are, in essence, the codes through which the collective unconscious communicates its wisdom to the individual.

The existence of archetypes within the collective unconscious implies that we are all connected to a shared field of awareness. This field transcends personal experience, drawing from a deeper source of knowledge that encompasses the entirety of human history. When we encounter archetypal symbols in dreams, art, or mythology, we are tapping into this collective reservoir, accessing insights and wisdom that are both personal and universal. These symbols resonate deeply because they are part of the core language of the psyche, a language that speaks to the essence of who we are.

Archetypes as codes of the collective unconscious offer us a pathway to self-understanding and transformation. By engaging with these symbols, we can access the wisdom of the collective, uncover hidden aspects of ourselves, and integrate the various parts of our psyche into a coherent whole. This process of integration is not only a personal journey but also a collective one, as each individual's exploration of archetypes contributes to the evolution of the collective consciousness.

The Role of Archetypes in Personal Growth and Transformation

Archetypes serve as guides on the journey of self-discovery, offering a framework for understanding the stages of personal growth and transformation. The Hero's Journey, for instance, is an archetypal pattern that represents the process of overcoming obstacles, facing inner and outer challenges, and emerging with a greater understanding of oneself. This journey is not limited to myth and fiction; it is a psychological and spiritual path that each individual undertakes, a symbolic journey that reflects the quest for meaning, purpose, and wholeness.

As we engage with archetypes in our own lives, they help us navigate the complexities of human experience. The archetype of the Shadow invites us to confront our fears, insecurities, and hidden desires, encouraging us to integrate these aspects of ourselves rather than repressing them. The archetype of the Sage or Wise Old Man embodies the pursuit of wisdom, reminding us of the importance of knowledge, insight, and self-reflection. These archetypes provide a symbolic map for personal development, guiding us through the stages of growth, conflict, and self-actualization.

The transformative power of archetypes lies in their ability to bring the unconscious into consciousness. By engaging with these symbols, we can uncover hidden aspects of ourselves, confront unresolved conflicts, and gain insights that lead to greater self-awareness. This process of integrating archetypal energies is a path to wholeness, a way of becoming more aligned with our true nature and with the universal forces that shape existence.

Archetypes as a Language of the Universe

Archetypes are not only psychological constructs; they are also a language through which the universe communicates its deeper truths. Just as physics reveals the underlying patterns of matter and energy, archetypes reveal the fundamental patterns of consciousness and existence. They represent the cosmic forces that drive creation, transformation, and destruction, embodying the cycles of life, death, and rebirth that permeate the natural world.

In this sense, archetypes can be seen as the codes of existence, the universal symbols that reflect the laws and principles that govern reality. The archetype of the Creator, for example, symbolizes the force of creation, the drive to bring something new into being. The archetype of the Destroyer represents the necessary force of destruction, the breaking down of old forms to make way for the new. These archetypes reflect the cyclical nature of existence, the balance between creation and destruction that sustains the universe.

When we view archetypes as a language of the universe, we recognize that they are not merely human inventions but reflections of the underlying structure of reality. They are the symbols through which existence expresses its truths, the codes that reveal the unity between the inner and outer worlds. By understanding and engaging with these archetypal patterns, we align ourselves with the rhythms of the universe, gaining insight into the fundamental principles that govern life.

Engaging with Archetypes: A Pathway to Wisdom

Engaging with archetypes is a pathway to wisdom, a process of aligning with the deeper currents of existence and accessing the collective wisdom of humanity. Through mythology, dreams, art, and personal reflection, we can connect with these universal symbols, exploring their meanings and integrating their energies into our lives. This engagement is not only an intellectual exercise but a transformative experience, a journey into the depths of the psyche and the mysteries of existence.

As we engage with archetypes, we begin to see patterns and connections that transcend the personal, revealing the ways in which our lives are woven into a larger tapestry. We come to understand that our personal struggles, desires, and aspirations are part of a universal story, a story that has been told and retold in countless forms across cultures and centuries. This understanding brings a sense of connection, a recognition that we are part of something greater than ourselves, that our lives are expressions of the same forces that shape the cosmos.

By embracing the language of archetypes, we cultivate a sense of unity with the world around us. We learn to see our own experiences as part of a larger pattern, a reflection of the timeless themes that define existence. This awareness deepens our understanding of ourselves and others, fostering empathy, compassion, and a sense of purpose. It invites us to approach life as a journey of discovery, a quest to uncover the hidden patterns that shape our reality.

Archetypes as a Bridge Between the Conscious and the Unconscious

Archetypes serve as a bridge between the conscious and the unconscious, linking the visible and invisible realms of the psyche. They bring the unconscious into awareness, allowing us to integrate parts of ourselves that were previously hidden or repressed. This integration is a process of healing, a way of becoming whole by embracing all aspects of the self. By engaging with archetypes, we gain access to the wisdom of the unconscious, uncovering insights that enrich our understanding of who we are.

This bridging role of archetypes extends beyond the individual, linking us to the collective unconscious and the universal field of consciousness. Through archetypes, we connect with the shared experiences, symbols, and wisdom of humanity, tapping into a source of knowledge that transcends the personal. This connection fosters a sense of unity, reminding us that we are part of a larger story, a collective journey of discovery and transformation.

In this way, archetypes are more than symbols; they are pathways to deeper awareness, guides that lead us to the hidden dimensions of reality. By understanding the language of archetypes, we open ourselves to a richer, more expansive view of existence, one that honors both the individuality of the self and the unity of the collective.

The Timeless Language of Archetypes

Archetypes are the timeless language of the soul, a symbolic code that speaks to the deepest aspects of human experience. They are the universal patterns that shape our lives, the invisible forces that guide us on our journeys of self-discovery and transformation. By engaging with this language, we gain insight into the nature of existence, the structure of reality, and the interconnectedness of all things.

In understanding and embracing archetypes, we step into a world of symbols, a world where meaning is layered, multidimensional, and profoundly resonant. This symbolic language connects us to the wisdom of the ages, to the shared experiences and insights of humanity, and to the deeper truths that underlie existence. Archetypes remind us that our lives are part of a larger story, a story that unfolds within each of us and across the cosmos.

The language of archetypes is a pathway to wisdom, a guide to understanding both the self and the universe. It invites us to see beyond the surface, to explore the depths of the psyche and the mysteries of existence. By embracing this language, we align ourselves with the codes of existence, gaining insight into the timeless patterns that shape reality and discovering our place within the grand tapestry of life.

Symbolism as the Universe's DNA

Symbols are more than mere representations; they are the fundamental language through which the universe communicates its essence, structure, and processes. Just as DNA carries the genetic information that shapes all living organisms, symbols encode the core patterns and principles that define reality. They are the universe's way of transmitting meaning, guiding the unfolding of existence across dimensions, consciousnesses, and realms. Symbolism, in this sense, operates as a kind of cosmic DNA—a universal code that structures not only matter and energy but also the depths of consciousness and the fabric of existence itself.

This view of symbols as the universe's DNA suggests that reality is constructed from patterns of meaning rather than random occurrences. These symbols manifest in forms as diverse as mathematical formulas, archetypal images, natural patterns, and even the architecture of galaxies. They reveal the underlying unity of existence, showing that all things are interconnected and share a common language. From the spirals in seashells to the fractals in clouds, from the Fibonacci sequence in flowers to the cosmic patterns seen in galaxies, symbolism is woven into every aspect of the universe. It is as though the universe itself is speaking, communicating its secrets through an intricate language of shapes, numbers, and archetypes.

The Language of Patterns: Symbols in Nature and the Cosmos

The universe reveals its language through the patterns that appear across nature and the cosmos. These patterns are symbols that echo the same underlying structures across vastly different scales, reflecting a fractal-like quality that unites the microcosm and the macrocosm. The spiral, for example, is a symbol that appears in galaxies, hurricanes, seashells, and even the structure of DNA. It is a universal shape that represents growth, expansion, and the unfolding of potential, a symbol that captures the dynamic, ever-evolving nature of existence.

These recurring patterns—fractals, spirals, symmetry, and the Golden Ratio—are not random; they are expressions of the mathematical language that underpins reality. Mathematics itself is a form of symbolic language, one that captures the essence of physical laws and natural phenomena. This language of numbers and patterns is the universe's way of encoding information, a symbolic DNA that governs the formation of matter, the evolution of life, and the behavior of energy. It suggests that beneath the diversity of forms lies a unity of structure, a common code that orchestrates the symphony of existence.

In this view, symbols in nature are more than aesthetically pleasing or scientifically interesting; they are windows into the mind of the cosmos. By studying these patterns, we gain insight into the fundamental principles that govern reality, principles that operate on all scales and across all domains of existence. Symbols in nature invite us to see the universe not as a collection of separate objects but as an interconnected whole, a single, living system that expresses itself through an elegant and consistent language of form and pattern.

Sacred Geometry: The Blueprint of Existence

Sacred geometry is a term used to describe certain geometric shapes and patterns that are believed to hold spiritual significance. These forms, such as the Flower of Life, Metatron's Cube, the Sri Yantra, and the Platonic solids, are considered to be the building blocks of reality. They represent the structural principles underlying creation, serving as templates through which the universe manifests its diversity. Sacred geometry is a symbolic language that captures the essence of form and function, a universal blueprint that reflects the harmony and order inherent in the cosmos.

Each shape in sacred geometry holds unique symbolic meanings. The Flower of Life, for instance, represents interconnectedness and the unity of all life, while the Vesica Piscis, formed by the intersection of two circles, symbolizes duality and the generative power of creation. The Platonic solids—tetrahedron, cube, octahedron, dodecahedron, and icosahedron—are seen as the fundamental forms from which all matter is structured, corresponding to the classical elements of fire, earth, air, ether, and water. These shapes are found in natural phenomena, from crystals and plants to the molecular structures of organic compounds, suggesting that sacred geometry is indeed a universal code that permeates all levels of existence.

Sacred geometry acts as the DNA of the universe by providing a template for creation and growth. It illustrates how complexity can emerge from simplicity, how intricate structures can arise from basic forms. By engaging with these symbols, we gain insight into the creative intelligence that guides the universe, the harmonious principles that allow life and matter to flourish. Sacred geometry reminds us that the universe is not chaotic but orderly, a grand design where every part reflects the whole, and every form contains within it the code of existence.

Archetypal Symbols as the Language of Consciousness

Just as sacred geometry and natural patterns serve as the structural DNA of the physical universe, archetypal symbols serve as the DNA of consciousness. These symbols—such as the Hero, the Shadow, the Mother, and the Trickster—are patterns within the psyche that shape human thought, behavior, and experience. They are universal symbols that reflect fundamental aspects of the human experience, codes that guide the evolution of consciousness across cultures and generations.

Archetypal symbols function as the inner blueprint of the mind, organizing our thoughts, emotions, and actions according to universal themes. The Hero archetype, for instance, represents the journey of self-discovery and transformation, a path that each individual must navigate. The Shadow archetype embodies the hidden, often repressed parts of the self, challenging us to confront our fears and integrate all aspects of who we are. These archetypes operate as symbolic codes that guide the development of the psyche, shaping the path of personal growth and spiritual evolution.

In this way, archetypes are the symbolic DNA of consciousness, patterns that reflect the structure of the human psyche and the journey of self-realization. They connect us to a collective memory, a shared language of symbols that transcends individual experience. By engaging with these symbols, we access a deeper level of self-understanding, aligning with the universal principles that govern the growth and expansion of consciousness. Archetypal symbols remind us that we are not isolated individuals but expressions of a collective awareness, participants in a shared journey toward wholeness.

Symbols as Bridges Between the Visible and the Invisible

Symbols serve as bridges between the visible and invisible realms, linking the physical world with the realms of meaning, intention, and consciousness. They are the medium through which the unseen influences the seen, where spiritual and mental energies find form in the material world. Symbols translate abstract concepts into tangible shapes, allowing us to engage with forces that are otherwise intangible and ineffable. Through symbols, the invisible structures of reality—the principles of creation, consciousness, and evolution—become accessible to human understanding.

These bridges between worlds are evident in religious symbols, ritual objects, and mystical diagrams, which are designed to convey spiritual truths. The mandala, for example, is a sacred symbol used in various spiritual traditions to represent the universe and the self. Its symmetrical design symbolizes unity, harmony, and the interconnectedness of all things. Similarly, the yin-yang symbol embodies the balance between opposites, illustrating the dual nature of existence and the integration of light and dark. These symbols provide a language for understanding the mysteries of existence, a way to access the realms that lie beyond ordinary perception.

In a broader sense, symbols as bridges remind us that reality is multidimensional, that the material world is only one aspect of existence. By engaging with symbols, we gain access to deeper layers of reality, connecting with the spiritual, psychological, and metaphysical dimensions of life. These symbols serve as guides, helping us navigate the unseen aspects of existence and align with the universal forces that shape our lives. They reveal that the universe is not only a physical construct but a field of meaning, a tapestry woven from both visible and invisible threads.

The Universe as a Living Symbolic System

The universe itself can be understood as a living symbolic system, a dynamic field where meaning and matter are inseparably intertwined. Every atom, molecule, and cell carries within it the symbolic code of existence, just as every thought, emotion, and idea is part of a larger tapestry of meaning. The universe communicates through this symbolic language, expressing its truths in forms that range from the microscopic to the cosmic. From the movement of planets to the formation of crystals, from the structure of DNA to the patterns of galaxies, everything is an expression of the same underlying code.

This view of the universe as a symbolic system invites us to see reality not as a collection of separate objects but as a unified, interconnected whole. It suggests that every part of the universe is a reflection of the whole, a microcosm that contains within it the essence of the macrocosm. This is the principle of "as above, so below," a concept found in many mystical and philosophical traditions. It implies that the patterns we observe in one realm, whether physical or metaphysical, are mirrored in others, revealing the inherent unity of existence.

By understanding the universe as a symbolic system, we gain a new perspective on the nature of reality. We see that everything is meaningful, that every form and structure is a manifestation of the underlying codes that govern existence. This awareness allows us to engage with life as a participant in a grand design, to recognize that we are part of a living, breathing cosmos that communicates through a language of symbols. It invites us to approach life with reverence, curiosity, and a sense of wonder, knowing that every symbol we encounter is a doorway into the mysteries of existence.

Living in Alignment with the Universe's Symbolic Language

To live in alignment with the universe's symbolic language is to become fluent in the codes that structure reality. It is to recognize that every experience, every encounter, and every pattern carries within it a message, a symbolic meaning that reflects the larger forces at play. By becoming aware of these symbols, we cultivate a deeper relationship with the universe, learning to interpret its signs and align ourselves with its rhythms.

This way of living encourages us to approach life as a dance with meaning, a journey of continuous discovery where every symbol is a guide and every pattern is a teacher. We learn to see beyond the surface, to engage with the world as a field of symbols that reveals the hidden order of existence. By aligning with the universe's symbolic language, we become attuned to the flow of life, moving in harmony with the forces that shape reality.

In embracing symbolism as the universe's DNA, we open ourselves to a greater understanding of both the world and ourselves. We begin to see that we are not separate from the cosmos but integral to its design, expressions of the same symbolic codes that structure all of existence. Through this awareness, we find our place within the grand tapestry of life, understanding that we are both participants in and expressions of a living, symbolic universe. This realization brings a sense of unity, purpose, and wonder, revealing that the journey of life is a journey of deciphering the universe's codes, a journey of engaging with the infinite wisdom encoded in the language of symbols.

Chapter Seven: The Symbiosis of Time and Timelessness

The Illusion of Temporal Progression

Time, as we experience it, flows in a linear fashion—a relentless progression from past to present to future. This perception of time gives structure to our lives, helping us make sense of events and instilling a sense of continuity in our experiences. Yet, quantum physics and ancient philosophical teachings alike suggest that time may not be as straightforward as it appears. Instead of an objective, linear force that marches ever forward, time could be an illusion, a construct created by consciousness to navigate reality. Beneath this illusion lies a more complex, layered experience of time, one that transcends the rigid confines of temporal progression and invites us into the realm of timelessness.

The idea that time is an illusion challenges one of our most fundamental assumptions about reality. It implies that the past, present, and future are not distinct, separate points but aspects of a single, unified field. In this view, all moments exist simultaneously, intertwined within a timeless whole. What we perceive as the passage of time may simply be a result of our consciousness moving through different states, creating the experience of sequential events. This illusion of temporal progression is a necessary construct for navigating life, yet it limits our perception by confining us to a narrow view of reality, one that sees existence as a series of isolated moments rather than an interconnected continuum.

Time as a Construct of Consciousness

One way to understand the illusion of temporal progression is to see time as a construct of consciousness, a mental framework that allows us to interpret and organize our experiences. Just as space provides a physical context for objects to exist and interact, time provides a mental context for events to unfold. This framework helps us make sense of change, causality, and growth, giving shape to our experiences and allowing us to function in a world that seems to move forward. Without this construct, life would appear chaotic and disorienting, as we would have no reference points by which to order events or understand cause and effect.

However, this temporal framework may be a limitation rather than a reflection of true reality. Quantum physics suggests that particles do not behave in a strictly linear fashion; instead, they exist in states of superposition, where multiple possibilities coexist. This quantum behavior hints at a universe where all potential states are present simultaneously, a timeless realm where change is not defined by linear progression but by shifting perspectives. In this sense, consciousness uses time as a lens through which it views reality, creating the experience of past, present, and future to navigate an otherwise boundless field of possibilities.

By viewing time as a construct of consciousness, we begin to understand that our experience of temporal progression is not an absolute truth but a subjective interpretation. It is a way of ordering the infinite potential of the universe, a mental model that enables us to experience growth, memory, and anticipation. This model, however, may limit our perception, preventing us from seeing the deeper unity of existence that lies beyond the illusion of sequential events.

Timelessness in the Quantum Realm

The quantum realm reveals a reality where the boundaries of time and space dissolve. In the world of subatomic particles, events do not unfold in a neat, linear order. Particles can exist in multiple states simultaneously, and their behavior is determined not by a fixed sequence of events but by probabilities and potentialities. Quantum entanglement further blurs the concept of temporal progression, as entangled particles can influence each other instantaneously, regardless of the distance or time separating them. This suggests that the quantum realm operates according to a principle of timelessness, a state where all points are interconnected in a way that defies linear time.

In this quantum view, time may be more like a tapestry than a river, with every moment interwoven with every other. Instead of moving along a timeline, consciousness may be capable of perceiving or experiencing different points on this tapestry simultaneously. Timelessness, in this sense, does not mean the absence of events but the absence of separation between them. It implies a state of unity where all possibilities coexist, where the boundaries between past, present, and future are fluid rather than fixed.

This quantum perspective on time challenges us to consider the possibility that the universe itself exists in a state of timelessness, a boundless field where every potential reality already exists. Our experience of linear time may be a localized phenomenon, a function of how our consciousness interacts with this quantum field. From this perspective, what we perceive as temporal progression is simply one way of interpreting the infinite complexity of the universe, a method of translating timelessness into a sequence of experiences.

The Eternal Now: A Glimpse into Timelessness

Many spiritual traditions teach the concept of the "eternal now," a state of awareness that exists beyond the past and future. This "now" is not a fleeting moment between what has been and what will be; rather, it is an expansive state of consciousness where all things exist simultaneously. In the eternal now, time ceases to move forward, and we experience life as a seamless whole. This state of timeless awareness allows us to connect with a deeper reality, a dimension of existence that is not bound by temporal constraints.

The experience of the eternal now reveals that time is, in many ways, a psychological construct. Our thoughts and memories anchor us to the past, while our hopes and fears project us into the future. By releasing these mental anchors, we can step into a state of presence that is timeless, where the mind is free from the illusion of progression. In this state, we become aware of the interconnectedness of all things, sensing that every experience, every person, every moment is part of a unified field.

The eternal now offers a glimpse into the nature of timelessness, inviting us to experience reality as it truly is—whole, complete, and beyond the constraints of linear time. By cultivating this awareness, we can transcend the limitations of temporal perception, moving into a state of consciousness that aligns with the timeless nature of the universe itself. This state does not deny the existence of time; rather, it reveals time as a flexible, fluid aspect of reality, one that consciousness can transcend by shifting its perspective.

Time and Timelessness: A Symbiotic Relationship

While timelessness reveals the illusory nature of linear progression, time remains a valuable aspect of human experience. The interplay between time and timelessness is a symbiotic relationship, one that allows us to navigate both the physical and metaphysical realms. Time provides structure, enabling us to engage in the world, to learn, grow, and create. It offers a way to experience change and evolution, to understand causality and anticipate the future. Timelessness, on the other hand, invites us into a state of unity and presence, allowing us to perceive reality beyond the limitations of sequence and separation.

In this sense, time and timelessness are not opposites but complementary aspects of existence. Time gives us the experience of progression, while timelessness reveals the interconnectedness of all things. Together, they create a balanced experience of reality, one that honors both the journey of individual growth and the unity of all that exists. By learning to move between time and timelessness, we can live more fully, embracing both the dynamism of change and the peace of presence.

This symbiotic relationship between time and timelessness invites us to cultivate a dual awareness, one that honors the flow of life while recognizing the underlying stillness. It encourages us to engage with the world without becoming bound by it, to experience each moment as part of a larger, timeless whole. By embracing both aspects of reality, we can navigate the complexities of existence with greater clarity, seeing time not as a constraint but as a vehicle for exploring the timeless nature of consciousness.

Beyond Temporal Illusions: A Path to Expanded Consciousness

The illusion of temporal progression, while useful, can limit our perception and understanding. By seeing time as a construct rather than an absolute truth, we open ourselves to new possibilities of awareness and insight. When we transcend the illusion of time, we move into a state of expanded consciousness, a perspective that perceives reality as an interconnected, timeless web. This shift in awareness allows us to experience life with greater presence, to release attachment to the past and anxiety about the future, and to engage with each moment as a part of the eternal now.

Living beyond temporal illusions does not mean rejecting the experience of time but rather embracing a broader view. It means recognizing that while we move through time, we are also part of a timeless field of consciousness. This awareness brings a sense of peace and freedom, as we understand that the journey of life is not bound by linear progression but is part of a greater whole, a dance between the dimensions of time and eternity.

By embracing this expanded consciousness, we align ourselves with the deeper structure of reality, a structure that includes both change and permanence, movement and stillness. We come to see that time is not a barrier but a pathway, a means of exploring the boundless nature of existence. Through this awareness, we step into the symbiosis of time and timelessness, experiencing life as both a journey and a state of presence, a process of growth and a timeless unfolding.

Embracing the Mystery of Time and Timelessness

The relationship between time and timelessness invites us to embrace the mystery of existence. Time, with all its movement and change, is a gift that allows us to experience the richness of life, to evolve, to create, and to connect. Timelessness, on the other hand, reveals the essence of reality, a stillness that lies beneath all movement, a unity that transcends separation. Together, they form the dual nature of existence, a symbiosis that reflects the infinite complexity of the universe.

Embracing this mystery means living with a sense of wonder, acknowledging that time is both real and illusory, both a tool and a veil. It means recognizing that while we experience life as a sequence of moments, each moment contains within it the whole of existence. This awareness allows us to live with greater depth, to see each instant as a doorway into the timeless, each experience as a part of the eternal.

The illusion of temporal progression serves as a reminder that reality is more than it appears, that beyond the flow of time lies a dimension of timelessness that we are all a part of. By embracing this duality, we open ourselves to a richer, more expansive view of existence, one that honors the journey of life while recognizing the unity of all that is. In this way, the symbiosis of time and timelessness becomes not only a concept to understand but a path to live, a way of experiencing the beauty, mystery, and infinite potential of the universe.

Time Loops, Synchronicities, and the Eternal Now

The experience of time is often one of linear progression—a steady march from past to present to future. Yet, certain phenomena challenge this orderly flow, hinting at a reality where time loops, synchronistic events, and a state of eternal presence exist alongside the conventional timeline. These experiences reveal the symbiosis of time and timelessness, showing that while we live within the framework of linear time, there is a deeper layer of reality where all moments exist simultaneously, interconnected in ways that transcend causality. Time loops, synchronicities, and the concept of the eternal now offer a glimpse into this hidden dimension, inviting us to explore the timeless aspects of existence that subtly shape our lives.

The Mystery of Time Loops

Time loops are moments when we feel that events are repeating or that we are somehow revisiting a past experience. This phenomenon can manifest as déjà vu, the uncanny sensation that a current moment has happened before, or as a feeling of cyclical patterns in life that seem to replay with slight variations. Time loops challenge the linear understanding of time, suggesting that certain events, relationships, or experiences are interconnected in a way that transcends forward progression. These loops often carry a sense of familiarity, as if we are returning to a place or moment that exists outside of chronological time.

The experience of time loops hints at the possibility that time is not a straightforward, one-directional river but a cyclical phenomenon, one where certain moments or events repeat due to unresolved themes, emotional patterns, or karmic cycles. In this view, time loops are an opportunity for growth, inviting us to revisit and reinterpret experiences from a higher perspective. By recognizing these loops, we can break free from repetitive patterns and evolve toward greater understanding. Rather than being trapped by time, we become aware of the ways in which certain lessons, challenges, or relationships recur until they are fully integrated and resolved.

Time loops can also reveal the interconnectedness of past, present, and future, suggesting that certain events echo across time, influencing each other in complex and subtle ways. In this sense, time is not linear but layered, a multidimensional web where the past, present, and future inform one another. This perspective invites us to live with greater awareness, recognizing that the choices we make in the present may resonate backward and forward through time, shaping both past experiences and future possibilities.

Synchronicities: Signs of a Timeless Order

Synchronicity is the experience of meaningful coincidences—moments when unrelated events align in a way that feels significant, as if they are part of a hidden, interconnected order. Coined by psychologist Carl Jung, synchronicity suggests that there is a deeper structure to reality, a timeless field where thoughts, emotions, and events are connected beyond the limits of cause and effect. Synchronicities often defy logical explanation, yet they carry a sense of purpose, guiding us toward insights, decisions, or actions that align with our inner journey.

The occurrence of synchronicities reveals that there is more to reality than meets the eye. These meaningful coincidences suggest that the universe operates within a holistic framework, where every part is connected to every other part, and where time and space are not barriers but threads in a unified tapestry. In moments of synchronicity, we glimpse this underlying order, experiencing reality not as a random sequence of events but as a living, intelligent system that communicates through symbols, patterns, and connections.

Synchronicities often feel like messages from the universe, reminders that we are on the right path or that we are not alone in our journey. They can appear as numbers, symbols, or encounters with people who share insights that resonate deeply. These events invite us to move beyond a purely rational understanding of life and to embrace a more intuitive, holistic perspective. By paying attention to synchronicities, we open ourselves to the possibility that the universe is guiding us, that we are participants in a larger story that transcends the linear passage of time.

Living in the Eternal Now

The concept of the eternal now is central to many spiritual teachings, which propose that time is an illusion and that true reality exists in a state of timeless presence. The eternal now is not a fleeting moment between past and future; rather, it is a state of consciousness where we experience life as a unified whole, free from the constraints of time. In this state, we become aware of a timeless dimension within us, a part of consciousness that exists beyond the mental constructs of past and future. This awareness of the eternal now brings a sense of peace, presence, and connection, as we realize that we are part of a vast, interconnected field where all moments exist simultaneously.

Living in the eternal now is about being fully present, experiencing life without projecting into the future or dwelling on the past. It is a state of mind where we release attachments to what was or what might be, allowing us to engage with the present moment in its fullness. In this state, time becomes irrelevant; we move beyond the illusion of progression and experience life as it truly is—a continuous flow of experience, a moment of presence that is both infinite and complete.

The eternal now does not deny the existence of time; instead, it transcends it, revealing a dimension of reality where all moments are one. This state of timelessness allows us to perceive life from a higher perspective, to see the interconnectedness of events, people, and experiences without being bound by the illusion of separation. By living in the eternal now, we cultivate a deeper awareness of our own consciousness, realizing that we are not separate from the flow of life but an integral part of it. This realization brings a sense of unity, freedom, and purpose, as we experience life not as a series of disconnected events but as a coherent, meaningful journey.

The Interplay of Time Loops, Synchronicities, and the Eternal Now

Time loops, synchronicities, and the eternal now are not separate phenomena but interconnected aspects of a larger reality that operates beyond linear time. Time loops remind us that certain patterns and lessons are cyclical, reappearing until they are understood and integrated.

Synchronicities show us that events are connected in a way that transcends cause and effect, reflecting a timeless order that unites all things. The eternal now invites us to live in a state of presence, to experience life as a continuous, unified flow that exists beyond the constructs of past and future.

These phenomena reveal that time is not an absolute reality but a flexible, multidimensional field where moments are interwoven, where past, present, and future influence each other in ways that defy logical explanation. By recognizing the interplay of time loops, synchronicities, and the eternal now, we gain a deeper understanding of the nature of reality, one that is not bound by linear progression but is shaped by the timeless nature of consciousness. This awareness allows us to navigate life with greater wisdom, intuition, and openness, as we learn to trust the guidance of synchronicities, to learn from recurring patterns, and to experience each moment as part of an eternal present.

Embracing a Timeless Perspective

To embrace a timeless perspective is to see life as an interconnected whole, where every moment holds meaning and every experience is part of a larger pattern. This perspective invites us to move beyond the constraints of linear thinking, to recognize that time is both real and illusory, both a helpful framework and a flexible, fluid dimension. By living with an awareness of time loops, synchronicities, and the eternal now, we become attuned to the deeper rhythms of existence, sensing the hidden connections that shape our lives and guide our journeys.

This timeless perspective brings a sense of freedom and peace, as we realize that we are not limited by the past or driven by the future. We become aware that every moment contains within it the potential for transformation, growth, and insight, that each instant is an opportunity to align with the flow of life and to experience the unity of existence. By embracing this timeless awareness, we cultivate a way of being that is both grounded and expansive, a way of living that honors the mystery and beauty of the present while recognizing that all moments are part of the eternal now.

In this state, time loops become opportunities for growth, synchronicities become guideposts, and the eternal now becomes our home. We step into a new way of experiencing life, one that is free from the limitations of sequential time and open to the infinite possibilities of a timeless existence. This awareness transforms our understanding of reality, revealing that we are not separate from the universe but part of a symbiotic relationship with it, a relationship where time and timelessness dance together in a harmonious flow. Through this dance, we find a deeper connection to ourselves, to each other, and to the boundless mystery of existence.

Chapter Eight: The Evolution of Thought Beyond Words

Transcending Language: Thought as Pure Energy

Language is one of humanity's most powerful tools, a means of conveying ideas, emotions, and experiences. Words structure our thoughts and allow us to communicate, helping us build shared understanding and collective knowledge. However, as we explore the deeper nature of consciousness, we begin to encounter the limits of language. Words can only go so far in expressing the vast, multidimensional experiences of the mind. Beyond language lies a realm of pure thought, a state where ideas, insights, and emotions exist as raw energy, unbound by the constraints of words. This realm of thought-as-energy represents a higher level of awareness, where communication is direct, intuitive, and resonant, transcending the limitations of verbal expression.

In this state, thought is not confined by grammar, syntax, or vocabulary. It flows freely as a dynamic energy, more akin to a current of feeling, intuition, and insight. This form of thought is experienced as an immediate knowing, a kind of direct cognition that bypasses words entirely. Here, ideas are not defined by their labels; rather, they are felt and understood holistically. This experience of thought as pure energy allows for a deeper connection to the essence of ideas, a way of perceiving that goes beyond intellectual understanding and reaches into the realm of the intuitive, the symbolic, and the ineffable.

The Limitations of Language in Expressing Consciousness

Language, as sophisticated as it is, inherently reduces complex ideas into simplified forms. Words are symbols that point to meaning, but they are not the meaning itself. They are abstractions that only approximate the experience or concept they represent. For example, the word "love" attempts to capture an entire spectrum of emotions, from affection to devotion to unconditional compassion, but it can never fully convey the experience of love itself. Similarly, words like "spirit," "unity," or "infinity" can only hint at realities that transcend human understanding. Language, then, acts as a filter, shaping and limiting how we experience and express consciousness.

The limitations of language become especially apparent when we try to articulate profound, transcendent, or mystical experiences. Those who have had such experiences often struggle to find words that do justice to what they felt or saw. The experience exists as a rich, multidimensional reality, but in the act of translating it into language, much of its depth and nuance is lost. Words flatten the experience, turning it into something linear and two-dimensional, stripping it of the immediacy and fullness with which it was perceived. This is why mystics and philosophers throughout history have spoken of "indescribable" truths and "ineffable" states, pointing to the realm of experience that lies beyond the reach of language.

By recognizing the limitations of language, we open ourselves to new ways of knowing and communicating. We begin to see that thought itself can transcend words, existing as a direct form of energy that communicates essence rather than description. This realization invites us to explore thought in its purest form, to engage with consciousness as an energetic field where ideas, emotions, and insights resonate directly, unmediated by the boundaries of language.

Thought as Energy: Direct Cognition and Intuitive Understanding

When thought transcends language, it becomes a form of energy—a direct, nonverbal experience of knowing. In this state, thought is not separate from the thinker; it is an intrinsic part of consciousness, an expression of awareness that arises from within and radiates outward. This direct cognition allows for a form of understanding that is immediate and holistic, bypassing the need for words or mental constructs. It is an intuitive knowing, a sense of understanding that is felt rather than articulated.

Direct cognition can occur in moments of deep meditation, creative inspiration, or heightened awareness, when the mind quiets, and we tap into a state of presence where thoughts arise as pure energy. In this state, knowledge is experienced as a resonance within consciousness, a vibration that conveys meaning without needing to be translated into language. This experience is often described as a "knowing without knowing how"—a recognition that feels complete and self-evident, as if the information were already part of us.

Thought as energy operates beyond the dualities that language imposes. It allows us to perceive reality not in fragments but as an interconnected whole, a web of relationships where each part reflects the entirety. This mode of thought aligns with the nature of quantum consciousness, where particles and waves exist as potentialities and probabilities rather than fixed entities. By experiencing thought as energy, we access a higher level of awareness, one that perceives the unity of existence and the seamless flow of consciousness that underlies all things.

Telepathy and Nonverbal Communication: Sharing Pure Thought

The experience of thought as energy opens up the possibility of telepathic communication, a form of connection where thoughts and emotions are shared directly between minds without the need for words. Telepathy has been reported in various cultures and traditions, often in states of deep empathy, shared intent, or heightened consciousness. In these moments, individuals seem to "tune in" to each other's thoughts or emotions, creating a resonance that allows for an intuitive, nonverbal exchange of information.

Telepathy challenges the conventional understanding of communication by suggesting that thought can exist independently of language, as a pure form of energy that resonates between individuals. In telepathic communication, ideas and emotions are not encoded into words and then decoded by the listener; rather, they are experienced directly, bypassing the translation process altogether. This form of communication reflects a deeper level of connection, one that operates on the frequency of shared consciousness and allows for an immediate and profound understanding.

Nonverbal communication also reveals the power of thought as energy. Our body language, facial expressions, and eye contact convey emotions and intentions in ways that words cannot. These subtle forms of expression are often more accurate reflections of our true feelings than the words we use. They are energetic signals, vibrations that communicate our state of mind and emotional energy to those around us. By becoming attuned to these forms of communication, we can perceive thought as a field of energy, a continuous flow that transcends verbal limitations and resonates directly within consciousness.

The Collective Mind: Tapping into a Field of Shared Thought

If thought can exist as pure energy, then it may also operate within a collective field, a shared realm of consciousness where individual minds are interconnected. This idea of a collective mind aligns with Carl Jung's concept of the collective unconscious—a shared repository of archetypes, symbols, and experiences that underlie individual consciousness. In this field, thoughts are not confined to individual minds but resonate across a network of shared awareness, allowing for the transmission of ideas, emotions, and insights on a collective level.

When we experience thought as energy, we tap into this collective field, sensing that our minds are part of a larger whole. This awareness brings a sense of unity, as we realize that our thoughts, emotions, and intentions contribute to and are influenced by the collective consciousness. Moments of inspiration, creativity, and intuition may arise from this field, as we access knowledge that seems to come from beyond ourselves. This form of shared thought suggests that consciousness is not isolated but is part of an interconnected web, a network of energy that transcends individual boundaries.

By engaging with the collective mind, we expand our perception, moving beyond the limitations of personal experience and tapping into the wisdom of the whole. Thought as energy allows us to experience this connection directly, to perceive ideas not as isolated concepts but as vibrations within a field of shared awareness. This connection to the collective consciousness deepens our understanding, reminding us that we are part of a larger tapestry of thought, emotion, and insight.

Transcending Language: The Future of Communication and Consciousness

As we evolve in our understanding of consciousness, the limitations of language become increasingly apparent. The future of communication may lie in the development of ways to express thought as pure energy, to share knowledge and experience directly without the need for verbal translation. This evolution could lead to a form of communication that is both more authentic and more profound, one that conveys the essence of ideas without the distortions and limitations of words.

Transcending language would allow for a new level of connection, one where we experience thoughts, emotions, and intentions as direct expressions of consciousness. This form of communication could enhance empathy, compassion, and understanding, as it would foster a sense of unity and interconnectedness. By perceiving thought as energy, we open ourselves to a way of knowing that is fluid, intuitive, and resonant, a way of experiencing reality that aligns with the inherent unity of existence.

This evolution of thought and communication suggests that consciousness itself is moving toward a more integrated, holistic state. By embracing thought as pure energy, we transcend the boundaries of individual minds and tap into a shared awareness that reflects the unity of all things. This shift in consciousness represents a profound transformation, a movement from separation to unity, from words to direct experience, from fragmentation to wholeness.

Living in Alignment with Thought as Energy

To live in alignment with thought as energy is to recognize that our minds are not isolated but are part of a dynamic field of consciousness. This awareness invites us to cultivate mindfulness, to become aware of the energetic quality of our thoughts, and to engage with the world from a place of presence and openness. By perceiving thought as energy, we become more attuned to the subtle vibrations within and around us, sensing the interconnected web of consciousness that unites all beings.

This way of living encourages us to communicate with greater authenticity, to express ourselves not only through words but through presence, intention, and energy. We learn to listen beyond words, to perceive the underlying essence of what others are communicating, and to respond with empathy and understanding. This shift in perception allows us to experience life more fully, to engage with the world in a way that honors the unity of all things.

In aligning with thought as energy, we open ourselves to a new way of experiencing consciousness—a way that transcends language and embraces the timeless, boundless nature of mind. This awareness brings a sense of peace, as we recognize that we are part of a larger field of thought, a collective consciousness that reflects the interconnectedness of existence. Through this alignment, we step into a deeper understanding of reality, experiencing thought not as a separate, limited process but as an expression of the infinite energy that flows through all things.

In this state, thought becomes a tool for unity rather than division, a means of connection rather than separation. We experience ourselves not as isolated thinkers but as participants in a shared field of awareness, resonating within a universe that communicates through a language of energy, intention, and pure consciousness.

The Role of Intuition in Quantum Thinking

Intuition is a subtle yet profound faculty, a form of knowing that arises without the need for logical reasoning or verbal articulation. In the realm of quantum thinking—where reality is understood as a web of interconnected probabilities, potentialities, and energies—intuition becomes an invaluable guide. Unlike analytical thought, which relies on sequential steps and fixed structures, intuition accesses a deeper level of consciousness, tapping into a direct understanding that feels immediate and holistic. In quantum thinking, intuition acts as a bridge between the known and the unknown, allowing us to perceive connections and patterns that transcend the limits of rational thought.

Quantum thinking requires a willingness to embrace uncertainty, to recognize that reality is not a static, predetermined structure but a fluid field of possibilities. This requires moving beyond the rigid boundaries of conventional logic and opening oneself to a more expansive, receptive way of knowing. Intuition, in this context, is not merely a fleeting feeling or hunch; it is a form of direct perception that aligns with the fundamental principles of quantum mechanics. By cultivating intuition, we gain access to insights that arise from the interconnected nature of reality, allowing us to navigate complexity, ambiguity, and paradox with a sense of ease and flow.

Intuition as Quantum Perception

In quantum physics, particles do not exist as isolated entities; they exist in states of superposition, where multiple possibilities coexist until observed. Similarly, intuition perceives multiple layers of reality simultaneously, accessing a form of "quantum perception" that allows us to understand complex situations without reducing them to linear cause-and-effect relationships. This quantum perception enables us to sense the whole without analyzing the parts, to feel the essence of a situation or idea without needing to dissect it.

Intuition operates in the same way, sensing patterns, connections, and energies that are not immediately visible but are nonetheless present. This capacity for direct perception aligns with the nature of quantum mechanics, where particles and waves exist as fields of potential that only collapse into fixed forms when observed. Intuition "reads" these fields, perceiving possibilities and resonances rather than fixed facts. In this sense, intuition is a quantum faculty, one that allows us to perceive reality as an interconnected whole, a web of relationships that cannot be fully understood through analysis alone.

By embracing intuition, we learn to engage with the world in a way that mirrors quantum perception. We move beyond binary thinking, embracing a mindset that is open, fluid, and receptive. This intuitive approach allows us to perceive subtle shifts in energy, to sense the underlying dynamics of situations, and to understand the essence of people, places, and ideas. Intuition, as quantum perception, connects us to the flow of life, attuning us to the rhythms, patterns, and possibilities that shape existence.

The Subconscious Mind as a Quantum Processor

The subconscious mind plays a critical role in intuition, processing vast amounts of information that the conscious mind cannot fully grasp. It acts as a quantum processor, integrating sensory input, memories, emotions, and even subtle energetic impressions to produce intuitive insights. While the conscious mind works linearly, analyzing one idea or piece of information at a time, the subconscious mind operates on a non-linear level, detecting patterns and connections that transcend logical analysis.

This ability to process information holistically allows the subconscious mind to make quantum leaps, accessing insights that feel immediate and self-evident.

The subconscious mind's quantum processing power allows it to "sense" the potential outcomes of different choices, accessing a broader field of information than the conscious mind. This is why intuition often feels like an inner knowing that emerges from nowhere—a spontaneous recognition of truth that is not based on sequential reasoning. By learning to trust this process, we allow our subconscious mind to guide us, opening ourselves to insights that bypass the limits of conscious thought.

In this way, the subconscious mind serves as an interface between the conscious self and the quantum field of possibilities. It bridges the gap between rational understanding and intuitive perception, enabling us to make decisions that resonate with our deeper knowledge. By tuning into the subconscious mind, we access a wellspring of intuition, a source of wisdom that is both personal and universal, and that connects us to the greater field of consciousness.

Embracing Non-Linear Thinking

Quantum thinking requires a shift from linear to non-linear thinking, a movement from fixed sequences to a dynamic awareness of interconnected patterns. Intuition thrives in non-linear thinking, as it is not confined by the rules of logic or sequence. Instead of following a strict line of thought, intuition "jumps" between ideas, sensing associations and resonances that may seem unrelated on the surface but are deeply connected on a fundamental level. This non-linear approach allows us to perceive reality as a complex web rather than a straightforward path, to understand that everything is interconnected in ways that defy conventional reasoning.

Non-linear thinking invites us to embrace the paradoxes and ambiguities that quantum mechanics reveals, to recognize that reality does not adhere to fixed rules but is fluid, flexible, and full of potential. In this state of awareness, intuition becomes a guiding force, helping us navigate the uncertainties and possibilities that quantum thinking brings. By allowing intuition to lead, we open ourselves to a new way of perceiving reality— one that is not bound by time, space, or causality, but that honors the interconnectedness of all things.

In practical terms, embracing non-linear thinking means releasing the need for immediate answers and trusting the unfolding process of discovery. It requires a willingness to be patient, to let insights arise naturally, and to accept that understanding may come in waves rather than a straight line. By cultivating this non-linear approach, we create a mental space that is receptive to intuition, a space where thought can unfold freely, guided by the subtle energy of awareness rather than the rigidity of logic.

Intuition and the Unified Field of Consciousness

The concept of a unified field—a field where all particles, waves, and forces are interconnected—aligns with the nature of intuition. Intuition seems to draw information from a universal source, a field of consciousness that exists beyond the boundaries of individual minds. In moments of intuitive insight, we may feel as though we are accessing a collective memory or tapping into knowledge that is not limited to personal experience. This experience aligns with the idea that consciousness itself is a field, a shared dimension where thoughts, emotions, and energies intermingle and influence each other.

Intuition, then, is a faculty that connects us to the unified field of consciousness, allowing us to perceive information that lies beyond the limits of the personal self. This connection to the field enables us to access insights, ideas, and inspirations that resonate with universal truths, guiding us in ways that are both deeply personal and profoundly interconnected. By embracing intuition, we open ourselves to the guidance of this unified field, aligning our awareness with the larger currents of consciousness that flow through all things.

This alignment with the unified field enhances our ability to understand reality holistically, to perceive the underlying patterns that connect people, events, and ideas. It enables us to sense the coherence of existence, to recognize that our thoughts and actions are part of a larger symphony of consciousness. In this state of intuitive awareness, we experience life not as a collection of isolated moments but as an interconnected flow, a tapestry woven from the energies of the quantum field.

Cultivating Quantum Intuition: Practices and Techniques

Developing intuition within the context of quantum thinking involves practices that cultivate receptivity, presence, and openness to the unknown. Meditation, for example, quiets the mind, allowing us to access deeper levels of consciousness where intuition naturally arises. In meditation, we learn to release attachment to specific thoughts or outcomes, creating a mental space where intuitive insights can emerge spontaneously. This state of relaxed awareness is ideal for connecting with the subconscious mind, which often holds the key to intuitive understanding.

Mindfulness practices also enhance intuition by grounding us in the present moment. When we are fully present, we are more attuned to subtle energies, sensations, and insights that arise within consciousness. By paying attention to these subtleties, we develop an awareness of the energetic dimension of thought, sensing intuitive impressions as they arise rather than dismissing them as random or irrelevant.

Journaling is another valuable tool for cultivating intuition. By recording our thoughts, feelings, and intuitive impressions, we create a dialogue with the subconscious mind, allowing insights to surface that we might otherwise overlook. Journaling also helps us identify patterns, themes, and recurring symbols in our lives, which can reveal underlying dynamics and intuitive messages.

Trusting our intuition is perhaps the most important practice of all. Intuition often speaks softly, through subtle feelings or impressions, and it requires a certain openness to recognize and act upon these signals. By learning to trust this inner guidance, we strengthen our connection to the quantum field of consciousness, allowing intuition to guide us with increasing clarity and depth. This trust enables us to navigate life with a sense of flow, recognizing that we are part of a larger reality that supports and guides us at every step.

Intuition as a Path to Quantum Awareness

The role of intuition in quantum thinking is not merely to provide answers or insights; it is to open us to a new way of perceiving reality. Intuition invites us to transcend the boundaries of rational thought, to experience consciousness as a field of energy, potential, and interconnection. By embracing intuition, we cultivate a quantum awareness that sees beyond the surface of things, sensing the invisible forces and energies that shape existence.

This quantum awareness brings a sense of harmony, as we realize that we are part of a dynamic, interconnected universe where every thought, action, and experience is meaningful. Intuition becomes our guide in this expanded state of consciousness, helping us navigate the complexities of life with a sense of purpose, clarity, and flow. By aligning with intuition, we open ourselves to the possibilities of the quantum field, allowing our lives to unfold in ways that reflect the unity, creativity, and boundless potential of existence.

In embracing intuition as a path to quantum awareness, we move beyond the limitations of language and logic, stepping into a realm of direct perception and holistic understanding. This shift in consciousness represents an evolution of thought, a movement from separation to unity, from analysis to intuition, from linear progression to a timeless, interconnected flow. Through intuition, we access the deeper truths of existence, perceiving reality as it truly is—a field of infinite potential, a living, breathing expression of consciousness in its most expansive form.

Chapter Nine: Bridging Dimensions with the Mind's Eye

Visualization and Manifestation: The Mind's Interaction with Multidimensionality

The mind's eye—the faculty of imagination and visualization—has a profound power that goes beyond mental imagery. Through visualization, the mind taps into multidimensional realms, enabling us to interact with reality in ways that transcend the physical senses. Visualization is more than just "seeing" with the mind; it is a means of communicating with deeper levels of consciousness, where intention, emotion, and energy converge to shape our experiences. In this way, visualization is not only a tool for mental clarity but also a bridge to the multidimensional nature of reality. When combined with intention and emotion, visualization becomes a powerful method for manifestation, allowing us to actively participate in creating our reality.

Manifestation is often discussed as the process by which thoughts become things, by which intentions influence reality in tangible ways. Quantum physics suggests that consciousness is an active participant in the structure of reality, capable of influencing matter at the subatomic level. Visualization, when directed with focused intention, engages this quantum principle, transforming thought energy into physical experience. Through visualization, we project our intentions into the field of potentialities, communicating with dimensions that lie beyond the physical and drawing from them the energies needed to bring our visions into form.

The Science of Visualization: Interacting with the Quantum Field

Quantum physics proposes that reality exists as a field of probabilities, where particles and waves exist in multiple states until observed. This quantum field of potentialities is shaped by observation and intention, suggesting that consciousness itself plays a role in determining which potential realities become actualized. Visualization acts as a form of conscious observation, directing the mind's focus toward specific outcomes and aligning one's energy with the desired reality. When we visualize, we "collapse" certain probabilities by holding a clear, focused image in our mind, influencing the quantum field to bring about the conditions necessary for manifestation.

Neuroscience also supports the power of visualization. Studies have shown that visualizing an action activates the same neural pathways as physically performing it. This phenomenon, known as mental rehearsal, enhances skill acquisition and reinforces neural connections, effectively preparing the brain and body for future experiences. In this way, visualization primes our mind and body for the realities we wish to create, aligning our inner state with our desired outcomes. By engaging with the quantum field through visualization, we not only alter our brain's chemistry and wiring but also communicate our intentions to the multidimensional layers of reality.

This process is most powerful when visualization is combined with intention and emotion. Emotion acts as an energetic amplifier, infusing our visualizations with a sense of reality and urgency. When we visualize with genuine feeling—whether it's gratitude, joy, or excitement—our thoughts resonate more strongly within the quantum field, enhancing the likelihood of manifestation. By consciously engaging in visualization with focused intention and emotion, we bridge the dimensions of thought and physical reality, aligning ourselves with the energy of creation.

The Role of the Subconscious Mind in Visualization and Manifestation

The subconscious mind plays a central role in the process of visualization and manifestation. It operates as a vast reservoir of memories, beliefs, and emotions, shaping our perceptions and influencing our interactions with the world. The subconscious mind is highly receptive to images, symbols, and emotions, making visualization an effective way to communicate with this deeper layer of consciousness. By visualizing specific outcomes, we program the subconscious mind to align with our goals, setting in motion the internal processes needed to bring them into reality.

The subconscious mind does not differentiate between real and imagined experiences; it responds to the emotional and sensory input we provide. This is why vividly imagining a situation can create physiological responses similar to those we would experience if the event were actually occurring. By visualizing our desires with sensory detail and emotional intensity, we "convince" the subconscious mind that our goals are already being realized. This alignment influences our behavior, attitudes, and even our interactions with others, creating a ripple effect that brings us closer to our desired reality.

Furthermore, the subconscious mind is closely connected to the quantum field, acting as an intermediary between our conscious desires and the multidimensional energies of reality. By programming the subconscious through visualization, we access this intermediary, creating a bridge between our inner intentions and the outer world. The subconscious mind, when aligned with our goals, directs our thoughts, emotions, and actions toward manifestation, enabling us to interact with the multidimensional aspects of reality in a tangible way.

Visualization as a Portal to Multidimensional Realms

Visualization allows us to explore dimensions beyond the physical, tapping into realms of consciousness that are not bound by time or space. When we visualize, we create a mental space where past, present, and future coexist, enabling us to connect with the timeless, infinite aspects of existence. In this space, our thoughts and intentions are not limited by the constraints of linear reality; instead, they operate within a multidimensional field, where possibilities are infinite and interconnected.

In many ways, visualization serves as a portal to the multidimensional self—a self that exists across different times, places, and potential realities. By visualizing our ideal future, we connect with a version of ourselves that has already achieved our goals, aligning with the energy of that reality and drawing it into our present experience. This connection with the multidimensional self empowers us to manifest with greater ease and confidence, as we tap into the awareness that our desired outcomes already exist within the field of potentialities.

Additionally, visualization can be used to access past experiences, allowing us to heal, reinterpret, or reframe events that may be affecting our present. By visualizing ourselves in a past situation and infusing it with understanding, forgiveness, or compassion, we alter our relationship with that memory, releasing emotional blockages and transforming our energy. This ability to interact with different timelines and possibilities highlights the multidimensional nature of visualization, revealing it as a tool for self-discovery, healing, and transformation.

Techniques for Effective Visualization and Manifestation

Effective visualization involves more than simply picturing a desired outcome; it requires a holistic approach that engages the mind, body, and emotions. Here are some key techniques for enhancing the power of visualization and aligning with multidimensionality:

1. **Sensory Detail**: When visualizing, engage all five senses to create a vivid, immersive experience. Imagine the sights, sounds, smells, textures, and tastes associated with your desired outcome. This sensory detail signals to the subconscious mind that the visualization is real, enhancing its effectiveness.

2. **Emotional Resonance**: Infuse your visualization with positive emotions, such as joy, gratitude, or excitement. Emotion is the language of the subconscious and the quantum field, amplifying the energy of your visualization and increasing its impact.

3. **Present Tense Affirmations**: Use affirmations in the present tense to reinforce the feeling that your desired outcome is already happening. Statements like "I am successful" or "I am loved" create a resonance that aligns with the reality you wish to manifest.

4. **Mental Rehearsal**: Visualize the steps needed to achieve your goal, rehearsing the actions and behaviors that will bring you closer to your vision. This process prepares the mind and body for real-life implementation, creating a sense of confidence and readiness.

5. **Daily Practice**: Consistency is key to effective visualization. Engage in daily visualization sessions, dedicating a few minutes

each day to focus on your goals with intention and emotion. This practice reinforces your intentions and aligns your subconscious mind with your desires.

6. **Letting Go**: After visualizing, release attachment to the outcome and trust that it will manifest in its own time and way. This act of letting go allows the quantum field to respond freely, without interference from anxiety or doubt.

By incorporating these techniques, you create a powerful framework for manifestation, one that integrates thought, emotion, and energy in a harmonious flow. Visualization, when practiced consistently and with intention, becomes a means of interacting with the multidimensional aspects of reality, aligning the mind, heart, and spirit with the energies needed to bring about desired changes.

Living in Alignment with Multidimensional Manifestation

Living in alignment with multidimensional manifestation involves embracing the knowledge that reality is fluid, responsive, and deeply interconnected. This awareness invites us to see life not as a series of fixed events but as a dynamic interplay of energies, potentials, and probabilities. By visualizing our desires and engaging with the quantum field, we become active participants in this process, shaping our experiences through conscious thought and intention.

This alignment encourages a mindset of openness, creativity, and trust. Rather than being confined by external circumstances, we recognize that our inner state—our thoughts, emotions, and beliefs—holds the key to our reality. By maintaining a clear vision, we continually align ourselves with the energy of our goals, attracting opportunities, insights, and experiences that resonate with our intentions.

Living in alignment with multidimensional manifestation also means embracing the unknown, allowing space for unexpected outcomes and possibilities. The quantum field operates in ways that are often beyond human comprehension, arranging events and synchronicities that bring about our desires in ways we could not have anticipated. By surrendering control and trusting the process, we open ourselves to a reality that is abundant, expansive, and full of potential.

Visualization and Manifestation as a Path to Self-Realization

The practice of visualization and manifestation is not only a means of achieving goals; it is a path to self-realization. By engaging with the multidimensional aspects of reality, we deepen our understanding of who we are, exploring the limitless potential of consciousness and the creative power within. Each act of visualization is a journey inward, an opportunity to align with our highest self and to manifest a reality that reflects our true essence.

Through visualization, we learn to harness the mind's interaction with multidimensionality, bridging the realms of thought and experience, intention and reality. This practice fosters a sense of empowerment and connection, as we recognize that we are co-creators within the vast, interconnected tapestry of existence. By aligning our thoughts with our highest intentions, we bring forth a reality that is not only fulfilling but also a reflection of our deepest values, dreams, and potential.

In this way, visualization and manifestation transcend personal desires and become a spiritual practice, a means of engaging with the universe in a conscious, intentional way. By cultivating this practice, we transform our inner world, harmonizing our mind, heart, and spirit with the multidimensional flow of existence. Through this alignment, we discover that the mind's eye is more than a faculty of imagination; it is a portal to the infinite, a gateway to the boundless potential of the quantum field, and a pathway to the realization of our highest self.

How the Third Eye Perceives Quantum Layers of Reality

The concept of the third eye, often associated with the pineal gland, appears in ancient spiritual traditions as a gateway to higher perception. The third eye is believed to enable a form of "inner vision," allowing us to perceive beyond the physical realm and access hidden dimensions of reality. From a quantum perspective, the third eye can be thought of as a faculty that attunes the mind to subtler layers of existence, allowing us to sense energies, vibrations, and patterns that lie beyond ordinary sensory perception. This inner eye provides insight into the quantum layers of reality, bridging the physical and metaphysical realms and enabling us to experience consciousness as a vast, interconnected field.

In many traditions, the third eye is considered the seat of intuition, wisdom, and spiritual sight. Unlike our physical eyes, which observe the external world, the third eye perceives the inner dimensions of reality, accessing information that is not limited by time or space. Through the third eye, we gain insight into the energetic structures that underlie the material world, perceiving reality as a multilayered field of potentialities, interconnected forces, and multidimensional experiences. This expanded vision reveals the quantum nature of reality, where particles, waves, and consciousness exist as a dynamic, fluid matrix.

The Pineal Gland: The Physical and Spiritual Eye

The third eye is often associated with the pineal gland, a small endocrine gland located in the center of the brain. The pineal gland has a long history of being linked to mystical experiences and altered states of consciousness. Ancient cultures regarded it as the "seat of the soul" and a bridge to higher realms. Modern science has shown that the pineal gland produces melatonin, which regulates our sleep-wake cycles, and may also produce trace amounts of DMT, a compound associated with profound spiritual and visionary experiences.

From a metaphysical perspective, the pineal gland is thought to function as a receptor for subtle energies, attuning us to frequencies that are otherwise beyond our awareness. This gland, when activated, allows us to access higher states of consciousness, facilitating the perception of quantum layers of reality. In this way, the pineal gland serves as a physical manifestation of the third eye, bridging the biological and spiritual aspects of human perception. It acts as an antenna for consciousness, receiving information from beyond the physical realm and allowing us to sense the quantum energies that shape our experience.

Through practices like meditation, visualization, and breathwork, we can stimulate the pineal gland, activating the third eye and enhancing our ability to perceive beyond the material world. This activation allows us to experience reality in a more expansive way, sensing the interconnectedness of all things and accessing insights that transcend the limitations of ordinary thought. By tuning into the third eye, we become more receptive to the quantum field, perceiving the underlying structures and energies that shape existence.

Perceiving Energetic Patterns and Vibrations

One of the primary functions of the third eye is to perceive energetic patterns and vibrations that are not visible to the physical eyes. In the quantum realm, everything exists as a field of energy, and all matter is composed of particles that vibrate at different frequencies. The third eye, as an energetic receptor, is sensitive to these vibrations, allowing us to sense the underlying energy fields that constitute reality. This perception reveals a world that is alive with movement and interconnectedness, where forms are temporary manifestations of deeper, vibrational energies.

Through the third eye, we can perceive the aura—the energetic field that surrounds living beings—as well as the subtle energies present in places, objects, and even thoughts. This expanded perception provides insight into the health, emotional state, and intentions of others, as we sense the energetic imprints that lie beneath the surface. By attuning ourselves to these vibrations, we gain a deeper understanding of the dynamics at play in any given situation, allowing us to respond with greater awareness and compassion.

This ability to perceive energy fields is akin to observing the quantum layers of reality, where particles are constantly shifting, merging, and transforming. In the same way that quantum particles exist as both waves and particles, the energies perceived through the third eye reveal a fluid, interconnected reality where everything is in constant motion. This perception of energy layers invites us to move beyond a fixed view of reality, opening ourselves to the multidimensional nature of existence.

Accessing Timeless Wisdom Through the Third Eye

The third eye is not only a tool for perceiving energy; it is also a gateway to wisdom that transcends the limitations of time. Many who activate their third eye report an enhanced ability to access insights that feel timeless, as though they are drawn from a deeper, universal source of knowledge. This wisdom often comes in the form of intuitive flashes, sudden insights, or profound understandings that arise without logical reasoning. The third eye connects us to the quantum field of consciousness, where all knowledge and experience coexist, allowing us to access insights that are not bound by linear progression.

This timeless wisdom reflects the quantum principle that all possibilities exist simultaneously within the field of potentialities. When we tap into this field through the third eye, we access layers of reality that hold information from the past, present, and future. This perception transcends our usual experience of time, enabling us to sense potential outcomes, anticipate future events, or gain insight into past experiences. In this way, the third eye functions as a portal to the timeless dimensions of consciousness, providing us with a deeper understanding of our lives and the interconnected nature of existence.

Through practices like meditation and visualization, we can strengthen our connection to this timeless wisdom, learning to trust the intuitive insights that arise from the quantum field. By cultivating this awareness, we begin to perceive life from a higher perspective, recognizing the interconnected patterns and cycles that shape our experiences. This expanded perception allows us to navigate life with greater clarity, purpose, and insight, as we align ourselves with the wisdom of the quantum field.

Intuitive Vision and Multidimensional Awareness

The third eye enhances our capacity for intuitive vision, enabling us to perceive multiple dimensions of reality simultaneously. This multidimensional awareness reveals that reality is not a single, linear experience but a complex field where various possibilities, energies, and timelines coexist. Through the third eye, we gain access to these layers of existence, allowing us to sense the interconnectedness of events, the resonance between people and places, and the underlying currents that influence our lives.

This intuitive vision allows us to perceive not only the physical aspects of reality but also the emotional, mental, and spiritual layers that coexist with it. For example, when we meet someone, the third eye may reveal aspects of their character or life path that are not immediately visible. This expanded perception helps us understand the deeper context of our interactions, as we sense the energetic and spiritual dimensions that inform each encounter. By attuning to these dimensions, we gain a more comprehensive understanding of ourselves and others, fostering empathy, compassion, and insight.

In this way, the third eye functions as a bridge to multidimensional awareness, revealing the quantum nature of reality, where each moment contains infinite potentialities. By perceiving these layers, we develop a deeper sense of purpose and alignment, as we recognize that our thoughts, emotions, and intentions influence not only our immediate environment but also the broader field of consciousness. This awareness empowers us to live more intentionally, creating a reality that reflects our highest vision and aspirations.

Cultivating the Third Eye: Practices for Expanded Perception

To activate and cultivate the third eye, we can engage in practices that enhance our sensitivity to subtle energies and deepen our awareness of the quantum field. Here are a few key techniques:

1. **Meditation**: Meditation quiets the mind, creating a receptive state that allows us to perceive beyond ordinary consciousness. Focusing on the third eye during meditation can help activate this center, enhancing our intuitive vision and opening us to multidimensional awareness.

2. **Visualization**: Visualize a point of light or energy at the center of your forehead, where the third eye is located. Imagine this light expanding, filling your mind with clarity, insight, and openness. This practice strengthens the third eye's receptivity, aligning your awareness with higher dimensions of consciousness.

3. **Breathwork**: Deep, rhythmic breathing helps calm the nervous system and balance the body's energy centers. By focusing on your breath and visualizing energy moving to the third eye, you can activate this center and enhance your perception of subtle energies.

4. **Mindful Observation**: Practice observing people, places, and situations without judgment. Pay attention to the feelings, impressions, and intuitive insights that arise. This practice cultivates the third eye's ability to perceive beyond surface appearances, allowing you to sense the deeper energies at play.

5. **Journaling Intuitive Impressions**: Record your intuitive insights, dreams, and visualizations. This practice reinforces the

connection to your third eye, helping you recognize patterns and develop confidence in your intuitive vision.

Through these practices, we strengthen the third eye's perception, aligning ourselves with the quantum layers of reality and enhancing our ability to navigate life with greater awareness.

Living with an Open Third Eye: Integrating Quantum Perception

Living with an open third eye means embracing a multidimensional perspective that perceives life as an interconnected field of energies, patterns, and potentials. This expanded perception allows us to move beyond the limitations of ordinary thought, experiencing reality as a dynamic, fluid matrix where all things are interwoven. By integrating this quantum perception into our daily lives, we cultivate a deeper sense of purpose, presence, and alignment with the universe.

An open third eye fosters a sense of unity with all beings, as we perceive the underlying connections that bind us. We become more attuned to the energies that influence our lives, gaining insight into the subtle forces that shape our experiences. This awareness allows us to navigate challenges with clarity, as we recognize the larger patterns at play and align ourselves with the flow of existence.

In this state, life becomes a journey of discovery, as we explore the vast, interconnected dimensions of consciousness that lie beyond the physical realm. We develop a sense of wonder, curiosity, and openness to the mysteries of existence, realizing that the world we see is only a fraction of what truly exists. By living with an open third eye, we embrace our role as co-creators within the quantum field, shaping our reality through intention, awareness, and connection to the infinite potential of the universe.

In this way, the third eye becomes more than a source of insight; it is a pathway to the highest realms of understanding, a doorway to the quantum layers of reality that reveal the unity of all things. Through the third eye, we glimpse the boundless nature of consciousness, experiencing ourselves as both observers and participants in the unfolding dance of existence. This expanded vision not only deepens our relationship with reality but also empowers us to live a life of purpose, compassion, and alignment with the timeless wisdom of the quantum field.

Chapter Ten: Philosophy of the Quantum Self

Purpose in a Self-Organizing Universe

In a universe where everything is interconnected, where particles influence each other across vast distances and where consciousness plays a role in shaping reality, the concept of purpose takes on new dimensions. Traditional views often see purpose as an external directive, a preordained mission given by some higher authority. However, in a self-organizing universe—one that is dynamic, interdependent, and constantly evolving—purpose emerges from within. It is not imposed; it is co-created through our interactions, intentions, and awareness. Purpose, in this context, is an intrinsic quality of existence, a force that arises naturally from the unfolding complexity and unity of the cosmos.

A self-organizing universe suggests that everything, from the smallest particle to the largest galaxy, exists within a field of continuous creation and transformation. The universe, as quantum physics reveals, is not a static machine but a living, evolving system, one where every element contributes to the whole. This perspective shifts our understanding of purpose from a linear, goal-oriented endeavor to a relational, participatory experience. Purpose becomes less about reaching a fixed destination and more about aligning with the dynamic flow of life, participating fully in the unfolding patterns of existence, and contributing our unique presence to the larger field of consciousness.

Purpose as Emergent, Not Assigned

In a self-organizing universe, purpose is not assigned by an external force but emerges naturally from the relationships between all things. This view aligns with the principles of emergence, where complex systems and patterns arise from simple interactions. Just as ecosystems evolve and adapt without a central blueprint, so too does purpose emerge from the interconnectedness of our lives. Each of us, as a unique expression of consciousness, has a role to play, a pattern to contribute, and this role is not rigidly defined but is fluid, evolving as we grow and adapt within the larger whole.

This emergent view of purpose suggests that we are co-creators with the universe, actively participating in the evolution of consciousness. Our purpose is not fixed; it is a dynamic expression of who we are in relation to the ever-changing field of existence. This means that purpose is both personal and universal, rooted in our individual experiences yet connected to the greater tapestry of life. As we align with this deeper understanding, we begin to see purpose as an unfolding journey, a process of discovery that is intimately tied to our own growth, relationships, and contributions to the world.

In this sense, purpose is less about achieving specific outcomes and more about engaging with life in a meaningful way. It is about embracing the present moment, responding to the needs and opportunities that arise, and contributing our unique gifts to the collective whole. By participating in the self-organizing flow of the universe, we fulfill our purpose not by following a predetermined path but by being fully present, aware, and connected to the greater field of consciousness.

The Quantum Self: Aligning with the Flow of the Universe

The quantum self—our true self in the context of a self-organizing, interconnected universe—understands that purpose is not separate from life itself. Purpose is woven into the very fabric of our being, a reflection of the underlying unity and intelligence that pervades existence. The quantum self recognizes that by aligning with the flow of the universe, we are naturally guided toward experiences, relationships, and actions that resonate with our deepest truth. This alignment with the universe is not a passive surrender but an active participation, a willingness to listen, respond, and adapt to the flow of life.

Aligning with the flow of the universe requires us to cultivate awareness, presence, and openness to change. It is a process of tuning into the subtle energies and patterns that shape our lives, trusting that we are part of a larger design that is intelligent, purposeful, and self-organizing. The quantum self embraces this flow, recognizing that purpose is not a fixed point but a journey of becoming, an evolving expression of our inner potential and our connection to the whole.

In practical terms, aligning with the flow of the universe means listening to our intuition, following our passions, and trusting the guidance that arises from within. It means releasing rigid expectations, allowing purpose to unfold organically, and embracing the mystery of life as it reveals itself moment by moment. By aligning with this flow, we become co-creators with the universe, participating in the self-organizing process that shapes existence and discovering a purpose that is both deeply personal and universally connected.

The Interdependence of Purpose and Collective Consciousness

In a self-organizing universe, purpose is not an isolated endeavor; it is inherently relational and interdependent. Just as each cell in the body contributes to the health of the whole organism, each individual's purpose is interconnected with the greater collective consciousness. Our actions, thoughts, and intentions have a ripple effect, influencing not only our own lives but also the lives of others and the evolution of consciousness itself. This interdependence reveals that our purpose is not only about self-fulfillment but also about contributing to the larger field of consciousness, enhancing the collective well-being and expansion of awareness.

When we view purpose through the lens of interdependence, we recognize that we are all part of a larger ecosystem of consciousness, where each individual plays a unique role. Our purpose becomes a contribution to the evolution of the collective, a way of participating in the expansion and deepening of awareness across humanity and the cosmos. This understanding fosters a sense of responsibility, as we realize that our choices impact the whole, and that by fulfilling our purpose, we contribute to the harmony and growth of the universe.

This perspective invites us to approach purpose with a spirit of service, understanding that our fulfillment is linked to the well-being of others and the planet. By aligning our purpose with the needs of the collective, we tap into a deeper, more expansive source of meaning. We experience purpose not as a solitary pursuit but as a shared journey, one that connects us to a vast, interconnected network of beings, energies, and dimensions. This collective sense of purpose enhances our awareness, inviting us to act with compassion, creativity, and integrity as we contribute to the self-organizing flow of the universe.

Purpose as an Evolutionary Drive

Purpose in a self-organizing universe is not only an individual experience; it is also an evolutionary force that drives the expansion and complexity of consciousness. Just as particles organize into atoms, atoms into molecules, and molecules into life, purpose serves as a catalyst for growth, integration, and transcendence. This evolutionary drive is not imposed from the outside but arises from within, as each being strives to express its unique potential and contribute to the greater whole. Purpose, in this sense, is a natural expression of the universe's inherent creativity, a force that fuels the emergence of new possibilities and the evolution of consciousness.

For the quantum self, purpose is experienced as an inner calling, a sense of alignment with the evolutionary impulse that pervades existence. This drive toward growth and self-expression is not separate from the universe; it is an intrinsic part of the self-organizing process that shapes reality. As we pursue our purpose, we are participating in this evolutionary drive, contributing to the unfolding complexity and harmony of the cosmos. Our purpose, then, is not static or fixed; it evolves as we do, expanding in scope and depth as we align with the larger patterns of the universe.

This evolutionary view of purpose encourages us to approach life with curiosity, openness, and a willingness to explore the unknown. It invites us to see purpose as a journey of discovery, a continuous process of learning, adapting, and transforming. By embracing this evolutionary perspective, we release the need for certainty and control, allowing purpose to emerge organically as we engage with life. This approach fosters a sense of freedom and possibility, as we recognize that purpose is not a destination but a path of growth, a way of participating in the unfolding story of the universe.

Living with Purpose in a Self-Organizing Universe

Living with purpose in a self-organizing universe means embracing a dynamic, co-creative relationship with life. It means recognizing that we are both guided by and participants in a larger flow of consciousness, a field of intelligence that is responsive to our intentions, actions, and awareness. Purpose in this context is not something we find; it is something we create through our interactions with the world, through our choices, and through our willingness to align with the deeper patterns of existence.

To live with purpose in a self-organizing universe is to cultivate a sense of trust and surrender, understanding that we are part of a larger design that is intelligent and benevolent. This approach invites us to listen deeply to our inner guidance, to follow the threads of curiosity, joy, and meaning that arise naturally within us. It is a way of living that honors both the individuality of the self and the interconnectedness of all things, allowing us to express our unique gifts while contributing to the collective whole.

By embracing this perspective, we discover that purpose is not a rigid, linear path but a dance of co-creation, a way of engaging with life that is open, responsive, and adaptive. We become more attuned to the rhythms and flows of the universe, sensing when to act, when to wait, and when to trust in the unfolding process. This alignment with the self-organizing nature of existence brings a sense of harmony, as we realize that our purpose is an integral part of the cosmos, a unique expression of the universe's ongoing evolution.

The Freedom of Purpose Beyond Predetermined Destiny

In a self-organizing universe, purpose is liberated from the constraints of predetermined destiny. It is not a fixed, unchanging path that we must follow but an open field of possibilities that we co-create with the universe. This freedom allows us to explore, experiment, and redefine our purpose as we grow and evolve. Rather than being bound by a single, unalterable mission, we are invited to adapt, to embrace change, and to respond to the unique circumstances of each moment.

This freedom of purpose encourages us to approach life with flexibility and creativity, knowing that our purpose can shift and expand as we do. It invites us to see purpose as a living, breathing force that is intimately connected to our awareness, intentions, and experiences. By releasing the need for certainty, we open ourselves to a purpose that is fluid, responsive, and aligned with the self-organizing flow of the universe.

In this state of freedom, purpose becomes a source of inspiration and joy, a way of participating in the dynamic unfolding of existence. We experience purpose not as a burden or obligation but as an invitation to explore, to create, and to express our unique essence within the greater whole. This liberated view of purpose brings a sense of fulfillment and peace, as we realize that we are not bound by a predetermined fate but are free to co-create our journey in partnership with the universe.

In embracing purpose as an emergent, evolutionary, and co-creative force, we step into a new understanding of ourselves and our place in the cosmos. We recognize that we are not separate from the universe but are active participants in its self-organizing flow, contributing to the unfolding complexity and harmony of existence. Through this awareness, we discover a purpose that is both profoundly personal and universally connected—a purpose that reflects the infinite creativity, intelligence, and unity of the quantum self within the self-organizing universe.

The Significance of Self-Knowledge within Infinite Complexity

In a universe that is vast, multidimensional, and infinitely complex, the journey of self-knowledge takes on a profound significance. At first glance, the quest to understand oneself may seem like a small endeavor within the immensity of existence. However, within the philosophy of the quantum self, self-knowledge is a gateway to understanding the universe itself. The deeper we journey into self-awareness, the more we attune ourselves to the interconnected fabric of reality, where the boundaries between self and cosmos begin to dissolve. Self-knowledge is not merely a personal pursuit; it is a path to engaging with the deeper truths of existence, a means of perceiving the infinite complexity of reality with clarity, presence, and purpose.

In the quantum worldview, where particles are entangled and consciousness is a co-creative force in shaping reality, self-knowledge becomes a powerful tool for aligning with the universe's flow. By understanding ourselves—our thoughts, emotions, patterns, and potential—we gain insight into the forces that shape our lives. This self-awareness allows us to see beyond our individual experiences, recognizing our participation in a vast, interconnected field of consciousness. Through the process of self-knowledge, we move beyond the illusion of separation, experiencing ourselves as both unique expressions of the universe and integral parts of the whole.

Self-Knowledge as a Key to the Quantum Field

In the quantum model, everything exists as a field of potentialities, where particles and energies are interconnected in ways that transcend time and space. Self-knowledge acts as a bridge to this quantum field, aligning our awareness with the underlying unity and complexity of existence. When we engage in the process of self-discovery, we open ourselves to the subtle dimensions of consciousness, tapping into insights, intuitions, and perspectives that reflect the quantum nature of reality. Self-knowledge allows us to perceive our place within the quantum field, to sense the interconnectedness of all things, and to attune ourselves to the infinite possibilities that life offers.

By cultivating self-knowledge, we begin to understand how our thoughts, emotions, and intentions interact with the quantum field, shaping our experiences and influencing the world around us. This understanding empowers us to live with greater intention, recognizing that our inner state influences the outer reality we experience. The more deeply we know ourselves, the more clearly we perceive the energetic patterns that guide our lives, allowing us to make choices that align with our highest potential and the harmony of the universe.

In this way, self-knowledge becomes a tool for navigating the quantum field, a means of understanding the deeper currents that shape our lives. It allows us to move beyond surface-level perceptions, accessing the layers of consciousness where the mysteries of existence reside. By attuning ourselves to the quantum field through self-knowledge, we align with the flow of life, experiencing a sense of unity, purpose, and interconnectedness with all that is.

The Mirror of Complexity: Self-Knowledge in a Multidimensional Universe

In a universe that is infinitely complex, self-knowledge reveals the ways in which we are both unique individuals and reflections of the larger whole. Just as a single drop of water contains the same elements as the vast ocean, each of us holds within us a reflection of the universe's complexity. The journey of self-discovery is, therefore, a journey into the structure of reality itself. By understanding the intricate patterns, desires, fears, and potentials within ourselves, we gain insight into the dynamics that shape the cosmos. We come to see that the complexity within us mirrors the complexity of the universe, revealing the interconnected web of energies, experiences, and relationships that define existence.

This perspective encourages us to approach self-knowledge with a sense of curiosity and reverence, as we recognize that our inner world is a microcosm of the universe. The layers of thought, emotion, and energy that make up our consciousness are not separate from the greater whole; they are expressions of the same forces that govern the stars, the galaxies, and the quantum field. By exploring these layers, we gain a deeper understanding of how complexity operates, both within ourselves and in the broader universe.

Self-knowledge, then, becomes a journey of exploring this complexity with openness and humility, embracing both the light and shadow within us as part of the universal dance. By doing so, we align ourselves with the self-organizing principles of the universe, recognizing that the process of understanding ourselves mirrors the process of understanding existence. This journey of self-knowledge transforms our perception, allowing us to see life not as a series of random events but as a coherent, interconnected whole.

The Transformative Power of Self-Reflection

Self-knowledge requires us to engage in self-reflection, an act of turning inward to examine our beliefs, patterns, and motivations. This process of introspection is transformative, as it allows us to move beyond conditioned responses and habitual ways of thinking, opening us to new perspectives and possibilities. In a quantum universe, where reality is shaped by consciousness, self-reflection becomes a means of altering the field of possibilities that we inhabit. By becoming aware of our own inner patterns, we change the way we interact with the world, influencing both our personal experiences and the larger field of consciousness.

Self-reflection also deepens our empathy, as it enables us to recognize the universal patterns and struggles that we share with others. As we come to understand our own fears, desires, and motivations, we become more compassionate toward the experiences of others, recognizing that we are all navigating the same complexities of existence. This empathy fosters a sense of unity, as we realize that the journey of self-knowledge is a shared experience, one that connects us to the greater whole.

The transformative power of self-reflection lies in its ability to reveal the ways in which we are both individual and universal, separate yet interconnected. By embracing this paradox, we gain a deeper understanding of ourselves and our place in the universe, aligning with the quantum nature of reality and the interconnectedness of all things. Through self-reflection, we move beyond the limitations of ego, experiencing a sense of expansion, unity, and purpose that transcends our individual identity.

Self-Knowledge as a Path to Authenticity and Empowerment

In the journey of self-knowledge, we come to recognize the layers of conditioning and beliefs that have shaped our sense of self. By peeling away these layers, we uncover our authentic essence, the core of who we are beyond societal expectations and external influences. This authenticity is not a fixed identity but a dynamic expression of our inner truth, a reflection of our unique perspective and connection to the universe. By embracing our authentic self, we align with our true purpose, experiencing a sense of empowerment and fulfillment that arises from living in harmony with our deepest values.

Authenticity, in this context, is not about fitting into a predefined role but about discovering the unique gifts, passions, and insights that we bring to the world. As we deepen our self-knowledge, we become more aware of these gifts, recognizing that they are expressions of the universe's infinite creativity. This awareness empowers us to live with intention, to express our truth, and to contribute to the collective consciousness in a way that is both meaningful and aligned with our highest potential.

Self-knowledge, therefore, becomes a path to empowerment, a means of realizing our innate potential and purpose. By knowing ourselves deeply, we gain the confidence to live authentically, free from the constraints of external validation or societal expectations. This sense of empowerment allows us to engage with life fully, embracing both its challenges and its opportunities as expressions of the quantum field. In this state of authenticity, we become co-creators with the universe, participating in the ongoing evolution of consciousness and the unfolding complexity of existence.

The Unity of Self and Cosmos: A Quantum Perspective

Self-knowledge, in the quantum perspective, reveals the unity between the self and the cosmos. The more deeply we understand ourselves, the more we perceive our connection to the greater whole. This unity is not merely a philosophical concept but a felt experience, a sense of belonging to a vast, interconnected web of consciousness. By exploring our inner world, we come to see that our thoughts, emotions, and intentions are not separate from the universe but are integral parts of the cosmic dance.

This awareness of unity transforms our understanding of purpose, as we realize that our individual journey of self-discovery is part of a larger, universal process. We come to see that the quest for self-knowledge is a microcosm of the universe's quest for self-awareness, an expression of the cosmos awakening to itself through each conscious being. This understanding fosters a sense of reverence, as we recognize that our lives are not isolated events but are intimately connected to the greater whole.

The unity of self and cosmos also brings a sense of peace, as we realize that we are part of a self-organizing universe that is intelligent, purposeful, and harmonious. This perspective allows us to release the need for control, trusting in the flow of existence and embracing the journey of self-knowledge as a path to alignment with the universal consciousness. In this state of unity, we experience a profound sense of purpose, as we recognize that our inner journey is a contribution to the evolution of the whole.

Embracing the Infinite Journey of Self-Knowledge

In a universe of infinite complexity, the journey of self-knowledge is never complete. Each layer of understanding reveals deeper mysteries, inviting us to continually explore, grow, and expand our awareness. This infinite journey is a gift, an opportunity to experience life as a process of ongoing discovery, a path of transformation that aligns us with the quantum field of potentialities. By embracing the journey of self-knowledge, we become active participants in the evolution of consciousness, contributing to the unfolding complexity and harmony of existence.

113

This journey also brings a sense of humility, as we recognize that self-knowledge is a path of lifelong learning. No matter how deeply we understand ourselves, there will always be new dimensions to explore, new insights to gain, and new ways of perceiving reality. This openness to growth allows us to engage with life with curiosity and wonder, seeing each moment as an opportunity to deepen our self-awareness and expand our consciousness.

Ultimately, the significance of self-knowledge within infinite complexity lies in its ability to connect us to the larger mystery of existence. Through self-discovery, we come to experience ourselves as both individual and universal, as unique expressions of the quantum field and integral parts of the whole. This awareness transforms our lives, allowing us to live with purpose, authenticity, and a profound sense of connection to the cosmos.

In embracing the journey of self-knowledge, we find our place within the self-organizing universe, aligning with the rhythms, patterns, and possibilities that shape existence. We discover that the path to understanding ourselves is the path to understanding the universe itself, a journey of infinite depth, beauty, and meaning that reveals the interconnected nature of all things. Through self-knowledge, we awaken to the unity of self and cosmos, experiencing life as a sacred dance within the boundless, ever-evolving field of consciousness.

Chapter Eleven: The Final Boundary – Surrender to the Unknown

Accepting the Unknown as the Final Frontier

In the exploration of consciousness, self, and the cosmos, we eventually encounter a boundary that even our most profound insights and advanced theories cannot cross. This boundary is the unknown—the vast, limitless expanse of reality that lies beyond human comprehension. No matter how much we uncover, there will always remain mysteries that defy explanation, realms that our minds cannot penetrate, and truths that elude even our most intuitive grasp. Accepting the unknown as the final frontier invites us to embrace humility and curiosity, to surrender to the mystery, and to see the unknown not as a barrier, but as an essential, ever-present aspect of existence.

The unknown is more than a limitation; it is a source of wonder, a realm of infinite potential that lies beyond our understanding. By acknowledging the unknown, we open ourselves to the possibility that reality is far more complex, beautiful, and interconnected than we can conceive. In this way, the unknown is not something to be feared or conquered, but a reminder of the boundless creativity of the universe. It is an invitation to let go of certainty, to find peace in ambiguity, and to live in awe of the mysteries that shape our existence.

The Role of Mystery in Expanding Consciousness

Mystery plays a vital role in the expansion of consciousness. It reminds us that life is not a closed system, neatly explained and defined, but an open field of exploration, discovery, and growth. When we encounter the unknown, we are forced to move beyond our familiar frameworks, to question our assumptions, and to stretch the boundaries of our perception. This encounter with mystery is both humbling and liberating, as it reveals the limitations of our knowledge while simultaneously expanding our awareness.

Mystery ignites our curiosity and inspires us to keep seeking, learning, and evolving. It reminds us that, no matter how much we know, there is always more to discover. This openness to mystery is essential for personal growth, as it allows us to approach life with a sense of wonder, adaptability, and resilience. By accepting the unknown, we cultivate a state of mind that is flexible and receptive, open to new insights, perspectives, and possibilities. This openness is the key to expanding consciousness, as it enables us to continually transcend our current understanding and move toward greater awareness.

In embracing mystery, we also develop a deeper appreciation for the interconnectedness of all things. The unknown reminds us that reality is a unified whole, a complex web of relationships that cannot be fully grasped through analysis alone. By surrendering to the mystery, we begin to experience ourselves as part of this interconnected field, recognizing that our lives are woven into the larger tapestry of existence. This awareness brings a sense of unity and belonging, as we realize that we are participants in the unfolding mystery of the universe.

Surrender as a Path to Wisdom

Surrendering to the unknown is a path to wisdom, one that requires us to let go of the need for control, certainty, and absolute answers. In a world that often values knowledge as power, surrender may seem counterintuitive. Yet, true wisdom arises not from possessing all the answers, but from recognizing the limits of our understanding and allowing space for the unknown. This surrender is an act of humility, a recognition that the universe is far vaster and more intricate than our minds can comprehend.

Surrendering to the unknown does not mean giving up on seeking knowledge or understanding; rather, it means approaching the search with a spirit of openness and acceptance. It is the willingness to hold questions without rushing to answers, to sit with ambiguity without forcing resolution, and to embrace paradox without seeking to simplify it. This

116

surrender allows us to access a deeper form of wisdom—one that is not rooted in certainty, but in the ability to flow with the ever-changing currents of life.

By surrendering to the unknown, we become more attuned to the subtle rhythms of the universe, sensing the patterns and possibilities that lie beyond our conscious awareness. This state of surrender cultivates a quiet, receptive mind, one that is open to insight, intuition, and inspiration. In this way, surrender becomes a pathway to wisdom, enabling us to engage with life from a place of presence, awareness, and trust in the unfolding of the universe.

The Paradox of Knowing and Not-Knowing

In the journey of self-discovery and exploration, we encounter a paradox: the more we know, the more we become aware of what we do not know. Each answer leads to new questions, each discovery opens doors to new mysteries. This paradox of knowing and not-knowing is a hallmark of true wisdom, as it reflects a balanced perspective that values both understanding and humility. To accept the unknown as the final frontier is to honor this paradox, recognizing that knowledge and mystery are two sides of the same coin.

This paradox challenges us to hold both certainty and uncertainty, to appreciate the beauty of what we have discovered while remaining open to the vastness of what we have yet to uncover. It is a reminder that knowledge is always evolving, that truth is multidimensional, and that reality is not confined to our current understanding. By embracing this paradox, we cultivate a flexible, adaptive mindset that allows us to navigate life's complexities with grace and resilience.

The paradox of knowing and not-knowing invites us to live in a state of dynamic equilibrium, balancing the desire for clarity with a reverence for mystery. It teaches us that certainty is not the ultimate goal; rather, it is the willingness to engage with the unknown, to dance with ambiguity, and to find meaning in the midst of uncertainty. This balanced perspective brings a sense of peace, as we realize that we do not need to have all the answers to live a fulfilling, purposeful life.

Finding Purpose and Meaning in the Unknown

For many, the unknown is a source of fear, a reminder of life's unpredictability and impermanence. Yet, when we shift our perspective, we can find purpose and meaning within the unknown. The unknown is where potential resides, where possibilities that have not yet taken shape wait to be born. By accepting the unknown, we open ourselves to new experiences, insights, and transformations, allowing life to unfold in ways that we could never have anticipated.

Finding purpose in the unknown requires us to embrace the journey, to see life as an evolving process rather than a fixed destination. It is the understanding that purpose is not a final answer but an ongoing exploration, a quest that deepens as we grow. This perspective invites us to approach life with curiosity, to trust that each moment holds meaning, and to find fulfillment in the process of becoming. In this state of openness, we discover that purpose is not something we impose on life, but something we co-create with the universe as we engage with the mystery.

The unknown also teaches us the value of trust. By surrendering to what we cannot control or predict, we develop a sense of trust in the universe, a belief that there is a larger intelligence guiding our path. This trust allows us to navigate uncertainty with confidence, knowing that we are part of a greater whole, a cosmic dance that is unfolding in perfect harmony. By finding purpose in the unknown, we experience life as a journey of discovery, one that is rich with meaning, growth, and wonder.

Embracing the Unknown as a Spiritual Practice

Accepting the unknown as the final frontier is a profound spiritual practice, one that brings us closer to the essence of who we are and the nature of existence itself. This practice requires us to release the need for control, to quiet the mind, and to cultivate a deep sense of presence. In this state, we begin to experience the unknown not as an absence, but as a presence—a boundless, living field of potential that invites us to explore, learn, and evolve.

Embracing the unknown as a spiritual practice transforms our relationship with life. We learn to approach each moment with an open heart, seeing challenges as opportunities for growth and setbacks as lessons in resilience. This perspective allows us to move through life with grace, accepting both joy and sorrow as part of the mystery. We become less attached to outcomes, more at peace with change, and more willing to trust in the flow of existence.

This spiritual practice also deepens our connection to the universe. As we surrender to the unknown, we align ourselves with the greater intelligence that shapes existence. We come to see ourselves as participants in a larger process, a self-organizing field that is guided by a wisdom beyond our comprehension. This awareness brings a sense of unity, as we recognize that we are part of the mystery, an expression of the infinite creativity and consciousness that permeates all things.

The Freedom of Not Knowing

There is a profound freedom in accepting that we do not and cannot know everything. This freedom allows us to release the burden of certainty, to let go of the need for fixed answers, and to embrace life as a journey of exploration. In the absence of absolute knowledge, we find space for creativity, spontaneity, and openness. We become more adaptable, more resilient, and more willing to step into the unknown with courage and curiosity.

The freedom of not knowing also encourages us to live more fully in the present moment. When we are not focused on trying to control or predict the future, we can engage more deeply with the here and now, experiencing life as it is rather than as we think it should be. This presence brings a sense of peace and joy, as we realize that each moment holds its own beauty, wisdom, and meaning.

In this state of freedom, we are able to approach life with a sense of wonder, seeing each day as an opportunity to learn, grow, and evolve. We come to understand that life's greatest mysteries are not problems to be solved but experiences to be lived. By embracing the unknown, we free ourselves from the constraints of limited perception, opening ourselves to the boundless possibilities that lie within and beyond the present.

The Journey Beyond the Final Frontier

Accepting the unknown as the final frontier does not mark the end of our journey; rather, it is the beginning of a deeper, more expansive exploration. It is an invitation to step beyond the boundaries of the known, to explore the realms of intuition, inspiration, and higher consciousness.

This journey requires courage, humility, and a willingness to let go of the familiar. It is a journey into the heart of existence, where we encounter the mysteries that lie at the core of reality and the essence of our being.

In embracing the unknown, we come to realize that life itself is the ultimate mystery, a boundless field of potential that defies explanation. This realization transforms our understanding of ourselves and the universe, as we recognize that we are not separate from the mystery but are expressions of it. We are participants in the cosmic dance, co-creators in the unfolding story of existence.

The journey beyond the final frontier is a journey of awakening, a path that leads us to a deeper awareness of our true nature and our connection to the infinite. It is a journey that reveals the beauty, complexity, and wonder of life, a journey that invites us to live with purpose, presence, and a profound sense of peace. In accepting the unknown, we find freedom, wisdom, and fulfillment, experiencing life not as a series of answers but as a sacred mystery, an invitation to explore, to learn, and to become.

By embracing the unknown as the final frontier, we open ourselves to the boundless possibilities of existence, living with a sense of awe, humility, and gratitude. We come to understand that the mystery is not something to be solved but something to be experienced, a reminder that we are part of a universe that is both infinite and intimate, both unknowable and profoundly meaningful. This acceptance is the ultimate surrender, the final act of trust in the wisdom of the cosmos, and the doorway to a life of wonder, joy, and endless discovery.

Faith, Wonder, and the Conscious Embrace of Uncertainty

As we explore the limits of consciousness and the mysteries of existence, we inevitably confront the unknown—a vast, enigmatic space that defies explanation. In this encounter, three qualities become essential: faith, wonder, and a conscious embrace of uncertainty. These qualities help us navigate the final boundary, guiding us beyond the need for concrete answers and allowing us to dwell in the open, expansive state where questions are honored and mystery is welcomed. In this space, we find that uncertainty is not a threat to understanding but a doorway to profound insights, connections, and a deeper experience of reality.

Faith, in this context, is not blind belief or dogma; it is a trust in the intelligence of the universe, a recognition that there is an underlying order and wisdom within the unknown. Wonder keeps our sense of curiosity alive, reminding us that life is an unfolding mystery, filled with beauty and unexpected revelations. The conscious embrace of uncertainty, meanwhile, grounds us in humility, allowing us to recognize the limits of our knowledge while remaining open to the limitless possibilities that lie beyond. Together, these qualities enable us to surrender to the unknown with grace, resilience, and a sense of adventure.

Faith: Trusting the Intelligence of the Unknown

In a self-organizing, interconnected universe, faith is the quiet trust that life is unfolding as it should, even when we cannot understand it. Faith allows us to let go of the need for certainty, to release the compulsion to control or predict every outcome. Instead, we trust in the deeper intelligence of the cosmos, recognizing that we are part of a vast, dynamic process that is continually evolving. Faith in the context of the unknown is not about clinging to beliefs or absolutes; it is about opening to the larger flow of existence, allowing ourselves to be guided by the rhythms and patterns that shape the universe.

Faith also grounds us in the present moment. When we trust the intelligence of the unknown, we free ourselves from the constant search for answers, allowing us to engage more fully with the here and now. This faith encourages us to see life as a process of discovery, where each experience, each encounter, is an opportunity to learn and grow. By cultivating faith, we become more resilient in the face of challenges, more willing to take risks, and more open to the unexpected paths that life presents.

Faith in the unknown also fosters a sense of peace. It reminds us that we do not need to have all the answers, that we are not responsible for understanding or controlling everything. This peace comes from knowing that the universe is vast, wise, and purposeful, that we are held within a field of intelligence that is far greater than our individual selves. Through faith, we find the courage to surrender to the mystery, trusting that we are part of a larger story, one that is unfolding with beauty, complexity, and meaning.

Wonder: The Ever-Present Awe of Existence

Wonder is the quality that keeps our minds and hearts open to the infinite beauty and mystery of life. It is the ability to approach each moment with awe, to see the world as an endless source of inspiration, and to appreciate the miraculous nature of existence itself. In the face of the unknown, wonder becomes a powerful ally, reminding us that uncertainty is not something to fear but an invitation to explore. It is wonder that fuels our curiosity, that encourages us to ask questions, to look deeper, and to embrace the unknown with a sense of childlike openness.

Wonder also connects us to the present moment, helping us to experience life as it is, without the filters of judgment or expectation. In a state of wonder, we become aware of the intricate details and subtle beauty that often go unnoticed—the patterns in a leaf, the rhythm of a breeze, the brilliance of a sunset. This awareness deepens our connection to the world around us, allowing us to see that every aspect of life is part of a larger tapestry of existence.

Through wonder, we come to see that the unknown is not a void but a realm filled with potential, creativity, and meaning. Wonder invites us to experience life as a journey, not as a problem to be solved but as a mystery to be experienced. It allows us to live with a sense of gratitude and reverence, as we recognize that every moment holds within it a piece of the infinite, an aspect of the unknown that is both profound and beautiful. In this state of wonder, we find joy in the journey, a joy that is not dependent on answers but is rooted in the sheer wonder of being alive.

The Conscious Embrace of Uncertainty

To embrace uncertainty consciously is to accept that life's greatest mysteries may remain unanswered and to find peace in that realization. This conscious embrace of uncertainty is an act of humility, a recognition that the universe is far larger and more intricate than our minds can comprehend. By accepting uncertainty, we free ourselves from the limitations of rigid beliefs and narrow perspectives, opening to a broader, more expansive view of reality. This acceptance allows us to approach life with an open mind, willing to learn, adapt, and grow in response to the unknown.

The conscious embrace of uncertainty also teaches us the value of flexibility. When we release the need for certainty, we become more adaptable, more resilient, and more willing to navigate the twists and turns of life. We learn to trust our intuition, to follow our inner guidance, and to respond to life's challenges with creativity and presence. This flexibility enables us to move through life with grace, allowing us to flow with the changes and transitions that are an inevitable part of existence.

Embracing uncertainty is also a powerful way to cultivate inner peace. When we accept that we cannot control or predict everything, we release a tremendous amount of mental and emotional tension. We stop fighting against the natural flow of life, allowing ourselves to experience the present moment with openness and acceptance. This peace comes from knowing that we are part of a larger whole, a field of consciousness that is constantly evolving and transforming. By embracing uncertainty, we find freedom from the need for answers, a freedom that allows us to live more fully, more authentically, and more joyfully.

Living with Faith, Wonder, and Uncertainty

Living with faith, wonder, and a conscious embrace of uncertainty transforms our experience of life. It allows us to approach each moment with openness, curiosity, and trust, seeing the unknown not as a barrier but as an integral part of existence. This way of living invites us to find meaning in the journey itself, to appreciate the beauty of each moment, and to engage with life from a place of presence and awareness. By cultivating these qualities, we learn to navigate the mysteries of existence with grace, resilience, and joy.

This approach to life also deepens our connection to the universe. By embracing the unknown with faith, wonder, and acceptance, we open ourselves to the larger intelligence that guides existence. We come to see that we are not separate from the universe but are expressions of it, participants in the unfolding mystery of life. This awareness brings a sense of unity, a sense of belonging to something greater than ourselves. We experience life not as isolated individuals but as interconnected beings, part of a cosmic dance that is constantly evolving.

In this state, we find peace in the unknown, joy in the present moment, and fulfillment in the journey. We realize that the quest for answers is not the ultimate goal; rather, it is the willingness to engage with the mystery, to live with an open heart, and to trust in the flow of existence. Through faith, wonder, and the conscious embrace of uncertainty, we become more fully alive, more present, and more connected to the infinite potential that lies within and beyond us.

The Gift of Uncertainty: A New Way of Seeing

Uncertainty is often seen as a challenge, a source of fear or discomfort. But when we embrace it consciously, uncertainty becomes a gift, a doorway to a richer, more expansive experience of life. This new way of seeing allows us to let go of rigid expectations, to approach life with openness and adaptability, and to find meaning in the unfolding mystery of existence. Through uncertainty, we discover that life is not about reaching a final destination but about exploring the infinite possibilities that arise in each moment.

By accepting uncertainty as an inherent part of existence, we cultivate a deeper sense of compassion, empathy, and understanding. We recognize that we are all navigating the unknown, that each person we encounter is facing their own challenges, their own mysteries. This awareness fosters a sense of unity, as we see that we are all part of the same journey, participants in the same unfolding story. Through this perspective, we experience life as a shared adventure, one that is filled with beauty, complexity, and endless potential.

In the gift of uncertainty, we find the freedom to be ourselves, to explore, to create, and to grow. We realize that life is not about finding the "right" answers but about living fully, embracing the mystery, and experiencing the joy of discovery. Through this new way of seeing, we become more open, more resilient, and more willing to engage with life as it is, in all its beauty and complexity.

Beyond Knowing: The Path of Wonder and Faith

The path of wonder and faith leads us beyond the need for absolute knowledge, guiding us into a space where questions are celebrated, mysteries are honored, and the unknown is embraced with reverence. This path invites us to live with a sense of awe, to approach each day with curiosity, and to find joy in the unfolding journey. Through wonder and faith, we come to see that life's true meaning is not found in answers but in the experience of being alive, in the beauty of each moment, and in the connection we share with all of existence.

By following this path, we experience life as a sacred journey, one that is guided by a wisdom beyond our understanding. We learn to trust in the flow of life, to find peace in the midst of uncertainty, and to embrace the unknown as a source of endless inspiration. Through wonder and faith, we come to see ourselves as co-creators with the universe, participants in the grand adventure of existence. We realize that the journey itself is the destination, a path of growth, discovery, and transformation.

In the end, the conscious embrace of uncertainty, guided by faith and wonder, is the final act of surrender. It is the willingness to step beyond the known, to live with an open heart, and to trust in the wisdom of the cosmos. Through this surrender, we find the ultimate freedom, a freedom that allows us to experience life not as a series of answers but as a sacred mystery, a dance of consciousness, and a journey into the infinite.

Conclusion: The Eternal Spiral of Understanding

Wisdom Beyond Comprehension: Lessons from the
Boundless Journey

Wisdom Beyond Comprehension: Lessons from the Boundless Journey

As we reach the end of our exploration of quantum consciousness, we find that true understanding is not a destination but a path—a continuous journey that spirals ever outward, inviting us to embrace the mystery of existence with humility and wonder. In the pursuit of understanding, we encounter a paradox: the more we know, the more we become aware of the vast realms beyond our knowledge. This realization leads us to a wisdom that transcends comprehension, a deeper knowing that arises not from intellectual mastery but from an openness to the infinite complexity of life itself.

Wisdom beyond comprehension does not come from accumulating answers but from learning to live in harmony with the unknown. It is a wisdom that honors the mystery, recognizing that there are dimensions of reality, layers of consciousness, and realms of experience that defy explanation. This kind of wisdom is not confined to the mind; it is a quality of being, a state of presence that allows us to approach life with a sense of peace, acceptance, and awe. It is the recognition that we are participants in a boundless journey, one that continually expands our awareness and deepens our connection to the universe.

The Journey as the Destination

In the realm of quantum consciousness, the journey itself is the destination. There is no final point of arrival, no absolute truth to grasp, no end to the mysteries of existence. Each insight, each experience, and each revelation is a step along an endless path, a spiral of understanding that deepens and expands as we grow. This awareness transforms our relationship with knowledge, freeing us from the pressure to reach definitive answers and allowing us to savor the unfolding process of discovery.

The journey as the destination invites us to live fully in the present moment, to appreciate each phase of our exploration as a valuable part of the whole. It teaches us that every experience—whether it brings clarity or confusion, joy or challenge—is a necessary part of our growth. By embracing the journey, we become more open to the flow of life, recognizing that each twist and turn is an invitation to deepen our understanding and expand our consciousness.

This perspective also frees us from the need to achieve or prove ourselves. We come to see that our worth is not determined by what we know or accomplish but by our willingness to engage with life authentically, to explore the depths of our own being, and to connect with the world around us. The journey itself is a gift, a sacred dance that invites us to experience the beauty, complexity, and wonder of existence without needing to reach a final conclusion.

Embracing Paradox: The Unity of Opposites

One of the most profound lessons of the boundless journey is the ability to embrace paradox—the unity of opposites that exists within the fabric of reality. In the quantum realm, particles can exist in multiple states at once, embodying the duality of wave and particle, presence and absence, possibility and actuality. This paradox is not a contradiction to be resolved but a fundamental aspect of existence, one that reflects the interconnectedness of all things.

By embracing paradox, we learn to see beyond dualistic thinking, to recognize that life is not a series of opposites but a unified whole. Light and shadow, joy and sorrow, known and unknown—all are part of the same tapestry, each giving meaning and depth to the other. This unity of opposites teaches us to approach life with an open heart, allowing us to hold multiple truths at once and to find beauty in the complexity of existence.

127

Embracing paradox also allows us to live with greater balance and harmony. We learn to honor both our rational mind and our intuitive heart, to value both knowledge and mystery, to seek clarity while remaining open to the unknown. This balance brings a sense of peace, as we come to see that life does not require us to choose between opposites but to integrate them, to find our own unique expression within the vast, interwoven web of existence.

The Role of Humility in True Understanding

Humility is a foundational quality in the pursuit of wisdom beyond comprehension. It is the recognition that, no matter how much we know, there will always be realms of understanding that lie beyond our reach. This humility does not diminish our value or worth; rather, it grounds us in the reality of our human limitations, allowing us to approach the mysteries of existence with respect and reverence.

Humility teaches us to listen, to observe, and to learn from all aspects of life. It reminds us that every person, every experience, and every moment has something to teach us, if we are willing to be open. This openness allows us to grow continuously, as we learn to see life as a teacher, a mirror, and a guide. Through humility, we develop a deeper sense of empathy, recognizing that every being is navigating the same vast journey, each with their own unique insights and challenges.

In embracing humility, we free ourselves from the need to be right or to possess all the answers. We become more compassionate, more patient, and more willing to support others on their own paths. This humility allows us to approach life with a beginner's mind, open to new possibilities and perspectives, and willing to change as we evolve. Through this openness, we experience true understanding—not as a final answer but as an ever-expanding awareness that grows with each step of the journey.

The Interconnectedness of All Things

One of the most profound realizations on the boundless journey is the interconnectedness of all things. Quantum consciousness reveals that we are not isolated beings but integral parts of a vast, unified field of energy and consciousness. Every thought, action, and intention ripples through this field, influencing the whole and shaping the reality we experience.

This interconnectedness reminds us that we are participants in a shared journey, co-creators in the unfolding story of existence.

Understanding our interconnectedness brings a sense of responsibility, as we recognize that our choices impact not only ourselves but the entire web of life. This awareness encourages us to live with greater compassion, integrity, and mindfulness, as we align our actions with the well-being of the whole. It also brings a sense of unity, as we come to see ourselves as part of a larger, living system, one that is constantly evolving and expanding.

This realization of interconnectedness transforms our understanding of purpose. We come to see that our individual journey is not separate from the collective journey of humanity and the cosmos. Our growth, our insights, and our contributions are part of a larger process, a cosmic dance in which each of us plays a unique role. Through this awareness, we experience a sense of belonging and purpose that transcends individual desires, connecting us to the greater whole and to the infinite possibilities of existence.

The Infinite Potential of Consciousness

The journey into quantum consciousness reveals that consciousness itself is limitless, a field of infinite potential that holds within it all possibilities, all experiences, and all realities. This realization invites us to explore our own consciousness, to delve into the depths of our inner world, and to discover the vast potential that lies within us. It is a journey of self-discovery, one that reveals that our true nature is not confined to the boundaries of the physical self but extends into the boundless realms of consciousness.

The infinite potential of consciousness reminds us that we are creators, capable of shaping our reality through our thoughts, intentions, and awareness. It invites us to take responsibility for our inner state, to cultivate a mind and heart that resonate with the qualities we wish to experience in the world. Through this awareness, we realize that our consciousness is a doorway to transformation, a portal to the infinite, a space where we can connect with the universal flow of life.

In recognizing the infinite potential of consciousness, we come to see that our journey is not limited by our circumstances, beliefs, or past experiences. We are free to create, to grow, and to evolve in ways that are limited only by our imagination and openness. This awareness empowers us to live fully, to embrace the boundless journey with courage, creativity, and a sense of wonder.

The Legacy of the Boundless Journey: A Life of Purpose and Presence

The boundless journey of quantum consciousness is a path of awakening, one that leads us to a life of purpose, presence, and connection. By embracing the mysteries of existence, by honoring the unknown, and by cultivating a sense of wonder, humility, and faith, we transform our lives into a meaningful and fulfilling experience. We come to see that our purpose is not a fixed goal but an evolving expression of our true self, a way of contributing to the greater whole and participating in the unfolding story of the universe.

Living a life of purpose and presence means engaging with each moment fully, seeing each experience as an opportunity to learn, grow, and connect. It means living with a sense of gratitude, appreciating the beauty of the journey and the gifts of each day. This approach to life brings a sense of peace, as we realize that we are not alone but are part of a larger field of consciousness that supports and guides us.

The legacy of the boundless journey is not measured in achievements or accolades; it is found in the quality of our presence, in the depth of our compassion, and in the courage with which we embrace the unknown. It is a legacy of love, wisdom, and connection, one that touches the lives of others and contributes to the evolution of consciousness. Through this journey, we find a sense of fulfillment that is not dependent on external circumstances but is rooted in the boundless nature of our own being.

A Final Reflection: Embracing the Infinite Spiral of Understanding

As we conclude our exploration of quantum consciousness, we find ourselves on the edge of an infinite spiral of understanding. This spiral invites us to continue exploring, learning, and expanding, knowing that each step reveals new dimensions of reality and deeper levels of self-awareness. It is a journey that has no end, a path that continually opens to new mysteries, new insights, and new possibilities.

In embracing the infinite spiral of understanding, we accept that life is a boundless journey, one that is filled with beauty, complexity, and wonder. We recognize that our true purpose is not to reach a final answer but to engage fully with the experience of being alive, to explore the depths of our own consciousness, and to connect with the greater whole. Through this journey, we discover that we are both finite and infinite, both individual and universal, both seekers and participants in the grand mystery of existence.

May this journey bring you wisdom, joy, and a sense of wonder. May you find peace in the unknown, strength in your curiosity, and fulfillment in the boundless potential of your own consciousness. And may you, as part of the infinite spiral of understanding, live a life that reflects the beauty, mystery, and infinite possibilities of the quantum self.

An Invitation to Keep Seeking Beyond

As we conclude this journey through the realms of quantum consciousness, we find that the path of understanding is not linear but spirals infinitely, drawing us ever deeper into the mysteries of existence. This journey is not about arriving at final answers but about cultivating a way of being that is open, curious, and receptive to the boundless complexity of life. The closer we come to what we perceive as "answers," the more we see that these answers are but doorways to deeper questions, richer mysteries, and greater possibilities. In this way, the journey of understanding becomes a dance with the unknown, a continuous unfolding that invites us to keep seeking, exploring, and expanding our consciousness.

To keep seeking beyond is to embrace life as an adventure of endless discovery. It is to let go of the need for fixed truths and to open ourselves to the infinite nature of reality, where each moment holds the potential for new insights, connections, and transformations. This invitation is a call to live with a beginner's mind, one that is free from rigid expectations and alive with wonder. It encourages us to approach each experience, each encounter, and each moment with the openness to learn, to grow, and to see beyond the familiar.

Embracing the Journey with Openness and Curiosity

The journey of quantum consciousness teaches us that true understanding arises not from the accumulation of knowledge but from an open, curious engagement with life itself. This openness allows us to see beyond the surface of things, to perceive the interconnected web of energy, consciousness, and potential that underlies all of existence. By embracing the journey with curiosity, we become explorers of the unknown, seekers who are willing to venture beyond the boundaries of what is known and comfortable.

Curiosity is the spark that ignites our journey, the drive that propels us to ask questions, to challenge assumptions, and to expand our perception of reality. When we cultivate this curiosity, we find that life itself becomes a teacher, offering us insights and revelations in unexpected places. Each moment becomes an opportunity to deepen our understanding, to refine our awareness, and to connect with the greater whole. This openness transforms our lives, allowing us to experience reality not as a fixed structure but as a dynamic, evolving field of possibilities.

By embracing the journey with curiosity, we become participants in the unfolding mystery of existence. We learn to approach each question with humility, each answer with gratitude, and each moment with a sense of wonder. This way of being allows us to live with a sense of freedom and joy, knowing that life is a boundless journey of discovery, one that invites us to continually expand, evolve, and connect.

The Art of Asking Questions Without Demanding Answers

To seek beyond is to cultivate the art of asking questions without needing immediate answers. This approach honors the mystery of existence, recognizing that some questions may never have definitive answers. By embracing this openness, we allow ourselves to dwell in the unknown, to explore questions that stretch our understanding and challenge our perceptions. This practice teaches us to value the process of inquiry itself, to find meaning in the search rather than the solution.

When we ask questions without demanding answers, we open ourselves to deeper levels of insight, allowing our consciousness to expand and adapt. This openness creates space for intuition, creativity, and inspiration to arise, enabling us to perceive aspects of reality that lie beyond the limits of linear thought. In this way, the art of asking questions becomes a gateway to higher understanding, a practice that deepens our connection to the mysteries of existence.

This approach also cultivates patience and resilience, as we learn to live with uncertainty and to trust in the unfolding process of discovery. It encourages us to see each question as a stepping stone, a portal to new dimensions of awareness. By embracing the unknown with an open heart and a curious mind, we find that the journey itself is rich with insights, each question leading us to a deeper, more expansive understanding of life.

The Power of Wonder as a Guide

Wonder is a profound force that draws us beyond the ordinary, guiding us toward a deeper connection with the infinite. When we live with a sense of wonder, we become attuned to the subtle beauty and mystery that permeates every aspect of existence. Wonder keeps us grounded in the present moment, allowing us to experience life with freshness and vitality. It reminds us that reality is far greater and more mysterious than we can comprehend, inviting us to approach each day as an opportunity for discovery.

In the journey of seeking beyond, wonder is our guide, leading us to perceive the world not as a collection of fixed entities but as a dynamic, living field of possibilities. Wonder encourages us to explore the unseen realms, to reach beyond the familiar, and to embrace the extraordinary within the ordinary. It is a reminder that life is a gift, a precious opportunity to experience, to grow, and to connect with the larger cosmos.

By cultivating wonder, we open ourselves to the boundless potential of existence, finding beauty and meaning in places we might otherwise overlook. Wonder invites us to live with an open heart, to appreciate the miracles of life, and to approach each moment as a sacred experience. In this state, we are no longer merely observers of life; we are participants in a cosmic dance, co-creators in the ever-unfolding mystery of the universe.

An Invitation to Trust the Process of Becoming

The journey of seeking beyond is also a journey of becoming. As we explore the depths of consciousness, the mysteries of existence, and the infinite possibilities of life, we are transformed. Each insight, each experience, and each revelation shapes us, guiding us toward a deeper understanding of who we are and what it means to be alive. This journey invites us to trust in the process of becoming, to allow ourselves to grow, evolve, and expand without needing to know exactly where we are going.

To trust the process of becoming is to surrender to the flow of life, to recognize that we are part of a greater whole, an unfolding story that is both personal and universal. This trust allows us to move through life with grace, accepting both the joys and challenges as part of our growth. It frees us from the need for control, allowing us to experience life as it is, rather than as we think it should be.

In trusting the process, we come to see that each step of the journey has its own purpose and meaning. We recognize that even the moments of uncertainty, doubt, and difficulty are valuable, for they teach us resilience, courage, and faith. By embracing this trust, we find peace in the journey, knowing that we are exactly where we need to be, and that the path we are on is leading us toward greater understanding, connection, and fulfillment.

A Life Lived in Harmony with the Unknown

To keep seeking beyond is to live in harmony with the unknown, to find joy in the journey and peace in the mystery. It is an invitation to let go of rigid beliefs, to release the need for absolute certainty, and to embrace life as an ever-evolving experience. In living this way, we become more attuned to the rhythms and patterns of existence, finding ourselves in sync with the larger flow of consciousness.

A life lived in harmony with the unknown is a life filled with purpose and presence. It is a life that honors the journey as much as the destination, one that values growth, exploration, and connection above certainty or achievement. By embracing the unknown, we become more resilient, more compassionate, and more open to the beauty of each moment. We come to see that life is not a problem to be solved but a mystery to be lived, a journey that invites us to discover, to learn, and to awaken.

In this state of harmony, we find a sense of belonging, a recognition that we are part of a vast, interconnected whole. We experience life as a sacred dance, a continuous unfolding that reflects the boundless creativity, intelligence, and love of the universe. By living in harmony with the unknown, we honor our place in this dance, finding fulfillment not in final answers but in the experience of being fully alive, present, and engaged with the mystery of existence.

The Invitation to You, Dear Reader

As you close this book, remember that this is not an ending but an invitation—a call to keep seeking beyond, to continue exploring the realms of consciousness, and to embrace the infinite potential that lies within and around you. The journey of quantum consciousness is one that never truly ends, for each insight, each question, and each experience opens the door to new levels of understanding, new layers of reality, and new dimensions of self.

This invitation is a call to live with curiosity, with wonder, and with an openness to the unknown. It is a reminder that life is a precious, boundless journey, one that is rich with meaning, beauty, and connection. By accepting this invitation, you choose to engage with life fully, to explore the depths of your own being, and to connect with the greater whole.

May you continue to seek, to question, and to expand. May you find joy in the journey, peace in the mystery, and wisdom in the boundless complexity of existence. And may you, as a unique expression of the infinite, contribute to the ever-unfolding story of the universe, living a life that reflects the beauty, depth, and wonder of quantum consciousness.

In the end, the invitation to keep seeking beyond is an invitation to embrace the eternal spiral of understanding, to recognize that the journey itself is the destination, and to live each moment as a sacred expression of the boundless, ever-evolving dance of existence.

www.ingramcontent.com/pod-product-compliance
Lightning Source LLC
Chambersburg PA
CBHW071511220526
45472CB00003B/988